Holt Mathematics

Chapter 8 Resource Book

HOLT, RINEHART AND WINSTON
A Harcourt Education Company
Orlando • **Austin** • New York • San Diego • London

Printed in the United States of America

ISBN 0-03-078398-4

5 170 09 08

CONTENTS

Blackline Masters

Date ___________

Dear Family,

In this chapter, your child will learn how to find the perimeter and area of geometric figures such as rectangles, parallelograms, triangles, and trapezoids. Your child will also learn how to find the volume and surface area of pyramids, cones, prisms, and cylinders. Problems will have contexts such as woodworking, aviation, construction, entertainment, and computer aided design. In addition, your child will learn how to scale three-dimensional figures.

Perimeter is the distance around the outside of a figure. To find the perimeter of a figure, add the lengths of its sides.

Find the perimeter of the figure.

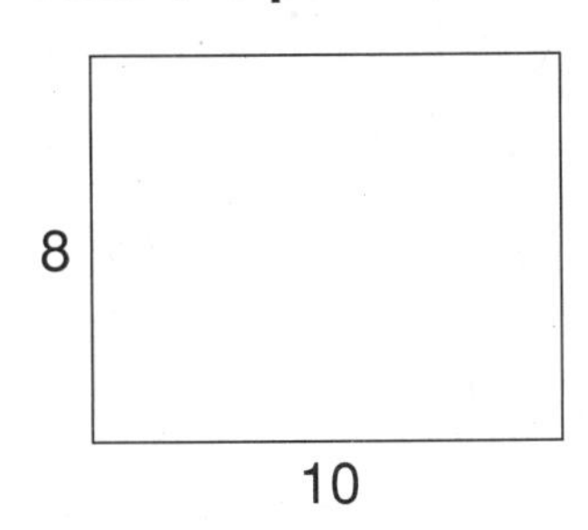

$P = 10 + 10 + 8 + 8$	*Add all side lengths.*
$P = 36$ units	
or $P = 2b + 2h$	*Perimeter of rectangle*
$P = 2(10) + 2(8)$	*Substitute 10 for b and 8 for h.*
$P = 20 + 16$	*Evaluate*
$P = 36$ units	

Area is the number of square units in a figure. The area of a rectangle or a parallelogram is the base length *b* times the height *h*.

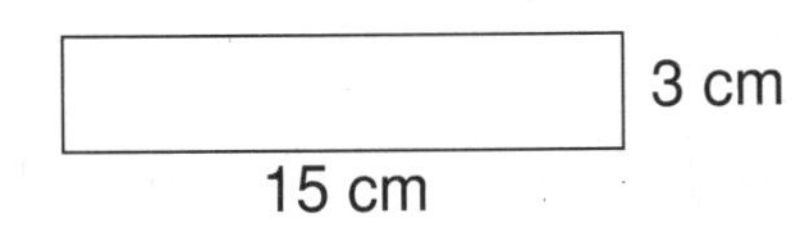

3 cm

15 cm

$A = b \times h$	$A = b \times h$	
$A = 15 \times 3$	$A = 15 \times 3$	Substitute 15 for the base
$A = 45\text{ cm}^2$	$A = 45\text{ cm}^2$	and 3 for the height.

Your child will learn how to find the **area** and **circumference** of a circle.

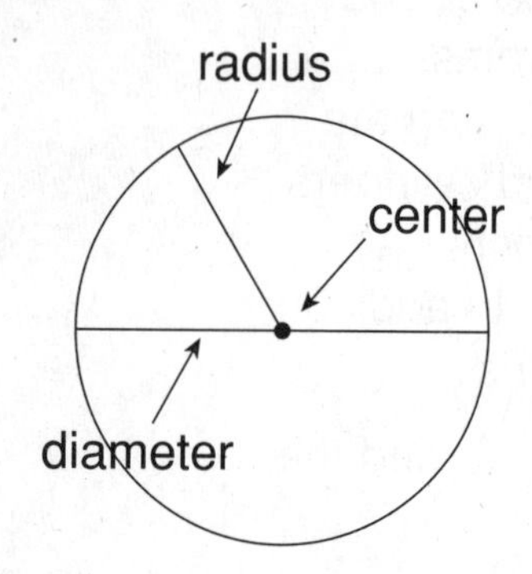

Circumference is the distance around a circle.
A **radius** connects the center to any point on the circle.
A **diameter** connects two points on the circle and passes through the center.

Circumference of a Circle		
Words	**Numbers**	**Formula**
The circumference C of a circle is π times the diameter d or π times twice the radius r.	3 6 $C = \pi(6)$ $C = \pi \cdot 2 \cdot 3$	$C = \pi d$ $C = 2\pi r$

Find the circumference of each circle both in terms of π and to the nearest tenth. Use 3.14 for π.

A circle with radius 5 cm

$c = 2\pi r$

$c = 2\pi(5)$

$c = 10\pi$ cm $\approx$ 31.4 cm

B circle with diameter 1.5 in.

$c = \pi d$

$c = \pi(1.5)$

$c = 1.5\pi$ in. $\approx$ 4.7 in.

For additional resources, visit go.hrw.com and enter the keyword MT7 Parent.

Name ______________________ Date ____________ Class ____________

LESSON 8-1

Practice A

Perimeter and Area of Rectangles and Parallelograms

Find the perimeter of each figure.

1.

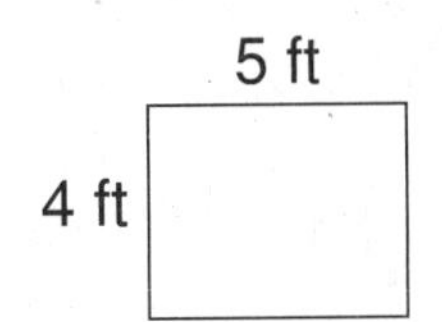

2.

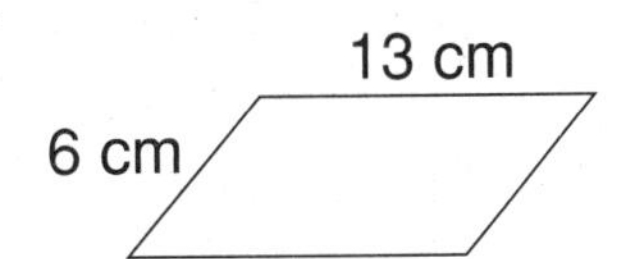

3.

______________ ______________ ______________

Graph and find the area of each figure with the given vertices.

4. (−3, 1), (2, 1), (2, −3), (−3, −3)

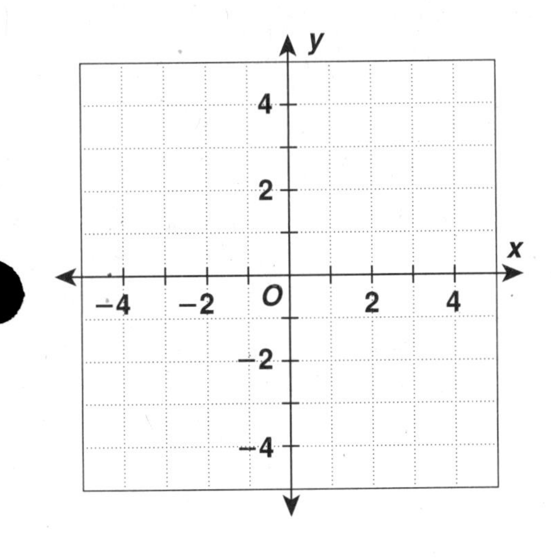

____

5. (−2, 3), (3, 3), (1, −1), (−4, −1)

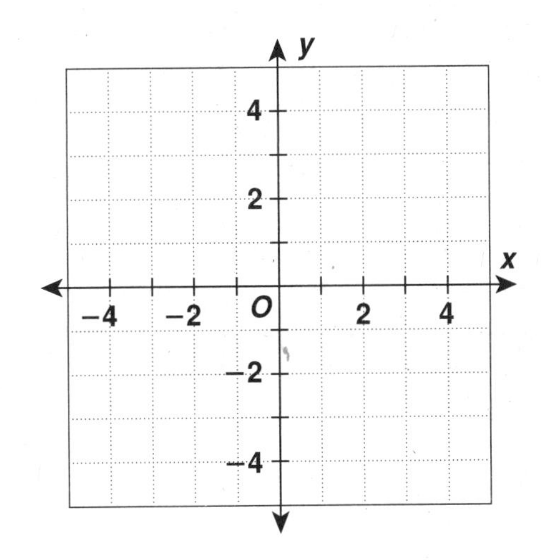

____

6. The Petersons plan to carpet their new family room. The carpet they have chosen costs $10 per square foot. They also need to install a pad underneath the carpet that costs $2 per square foot. How much will it cost the Petersons to install the carpet and the pad?

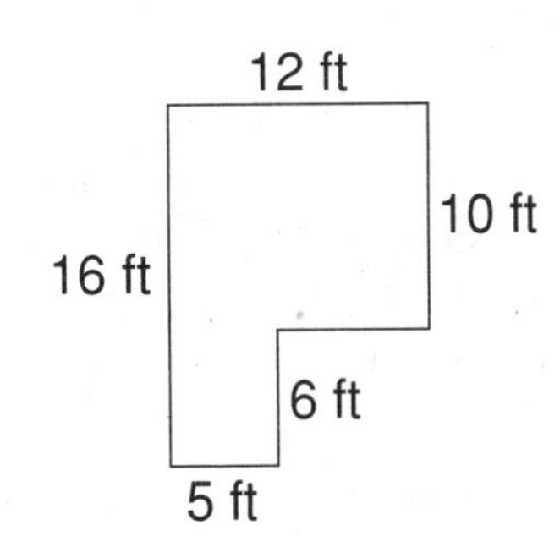

____

Name ______________________ Date __________ Class __________

LESSON 8-1

Practice B

Perimeter and Area of Rectangles and Parallelograms

Find the perimeter of each figure.

1.

2.

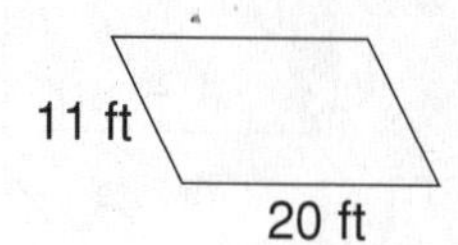

3.

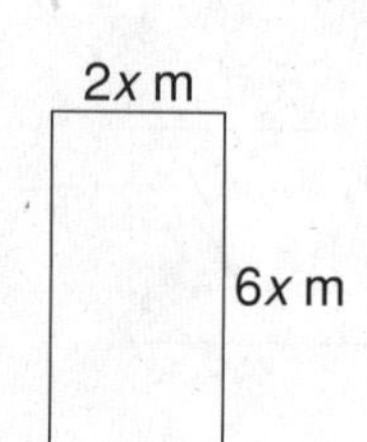

________________ ________________ ________________

Graph and find the area of each figure with the given vertices.

4. (−3, 4), (3, 4), (3, −4), (−3, −4)

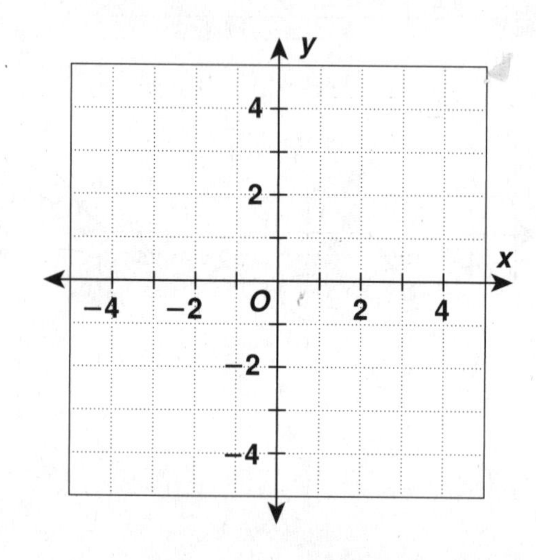

5. (−1, 3), (2, 3), (−1, −4), (−4, −4)

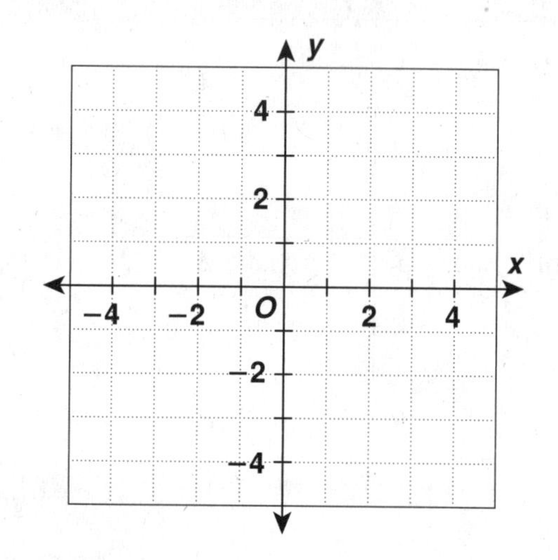

________________ ________________

6. Sloppi and Sons Painting Co. charges its customers $1.50 per square foot. How much would Sloppi and Sons charge to paint the rooms of this house if the walls in each room are 9 ft high?

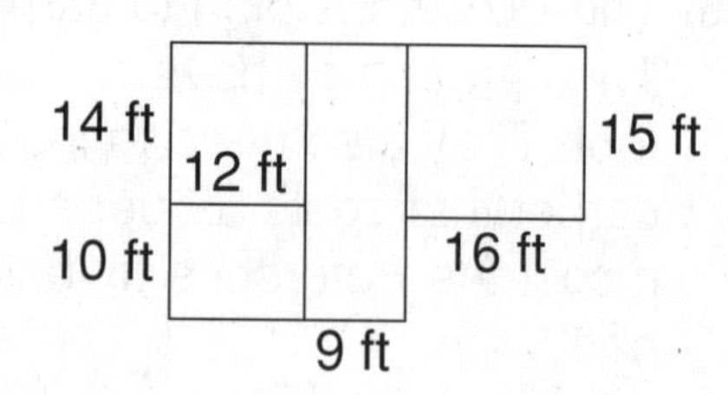

____

Name ____________________ Date __________ Class __________

LESSON 8-1

Practice C

Perimeter and Area of Rectangles and Parallelograms

1. Graph the figure with vertices and find the area of (−1, 3), (4, 3), (3, −4), and (−2, −4).

____

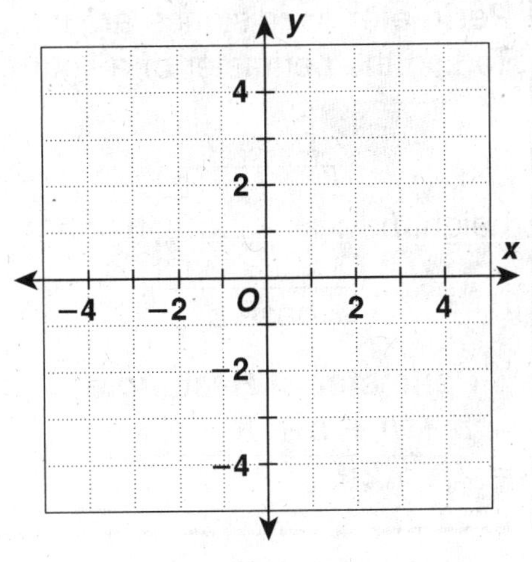

2. Graph the figure with vertices and find the area of (−2, 0), (2, 0), (0, −3), and (−4, −3).

____

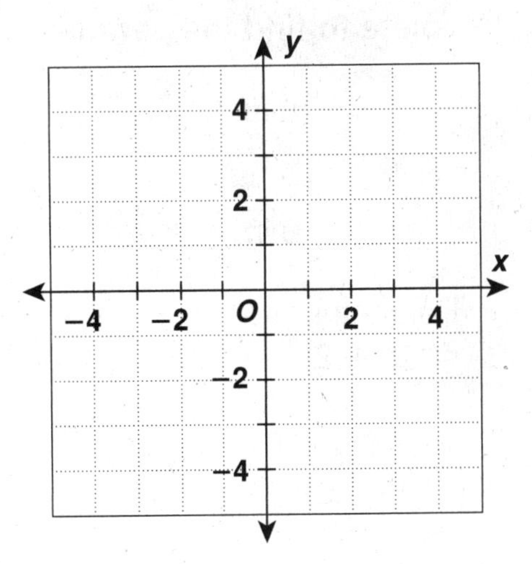

Find the perimeter and area of each figure.

3.

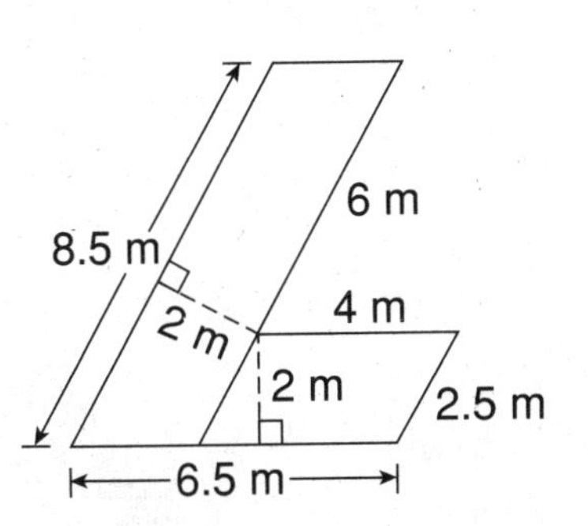

4.

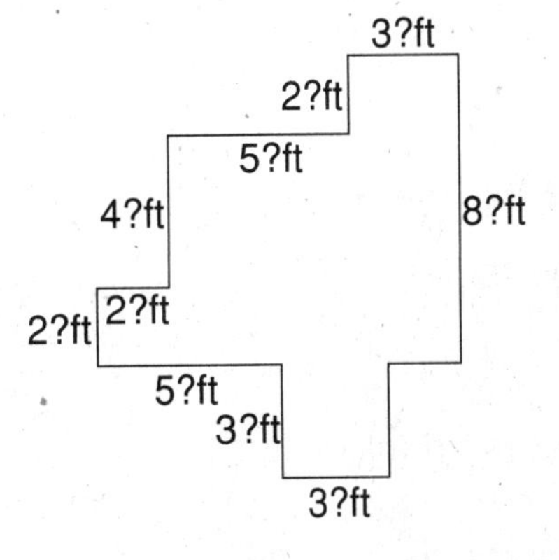

_______________________ _______________________

5. The Daughertys bought a piece of land that is 64 ft by 127 ft. They plan on building a two-story house that is 38 ft by 66 ft. How much land will remain after they build the house?

____

Name ______________________ Date ____________ Class ____________

LESSON 8-1

Reteach

Perimeter and Area of Rectangles and Parallelograms

Perimeter = distance around a figure.
To find the perimeter of a figure, add the lengths of all its sides.

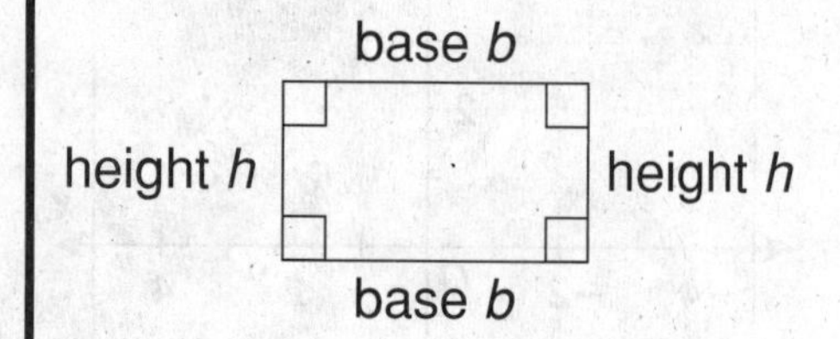

base b
side s
side s
base b

Perimeter of Rectangle
$= b + h + b + h$
$= \mathbf{2b + 2h}$

Perimeter of Parallelogram
$= b + s + b + s$
$= \mathbf{2b + 2s}$

Complete to find the perimeter of each figure.

1.

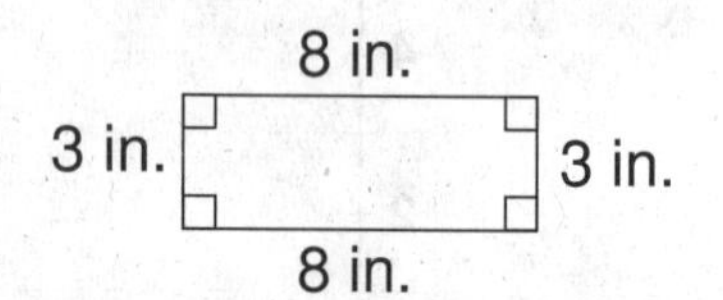

Perimeter of rectangle
$= 2b + 2h$

= 2(____) + 2(____)

= ____ + ____

= ____ in.

2.

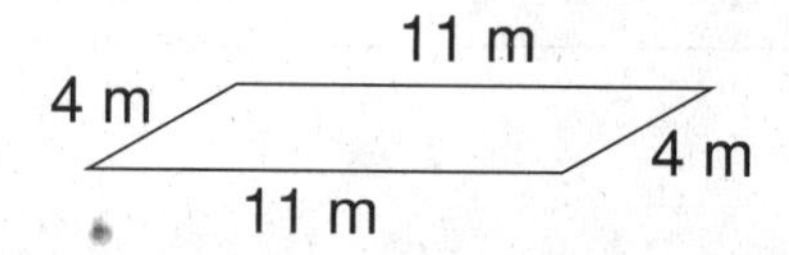

Perimeter of parallelogram
$= 2b + 2s$

= 2(____) + 2(____)

= ____ + ____

= ____ m

Find the perimeter of each.

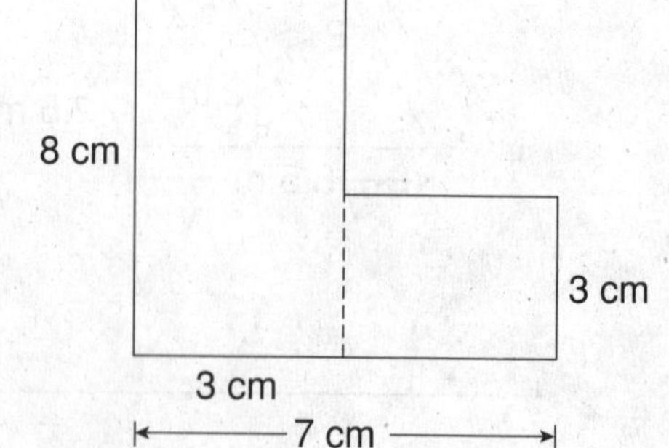

3. Large rectangle

P = ____ + ____ + ____ + ____ = ____

4. Small rectangle

P = ____ + ____ + ____ + ____ = ____

5. The combined rectangles as shown in the figure.

P = ____ + ____ + ____ + ____ + ____ + ____ = ____

Name ____________________ Date __________ Class __________

LESSON **8-1**

Reteach

Perimeter and Area of Rectangles and Parallelograms (cont.)

Area = number of square units contained inside a figure.

3

4

The rectangle contains 12 square units.

Area of rectangle = 4 × 3 = 12 units^2

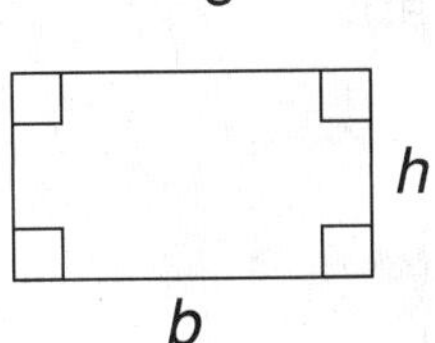

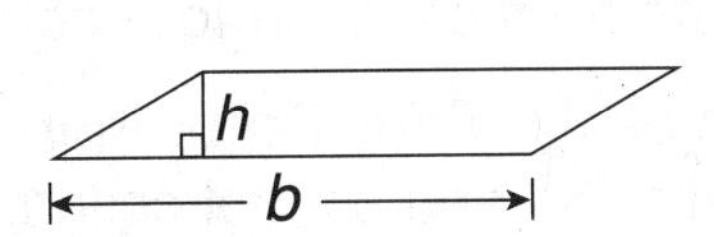

Area of Rectangle = $b \times h$ **Area of Parallelogram = $b \times h$**

Complete to find the area of each figure.

6.

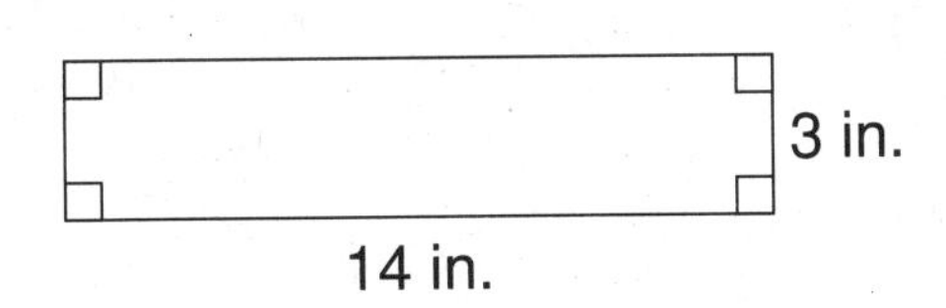

Area of rectangle
= $b \times h$

= ______ × ______ = ______ in^2

7.

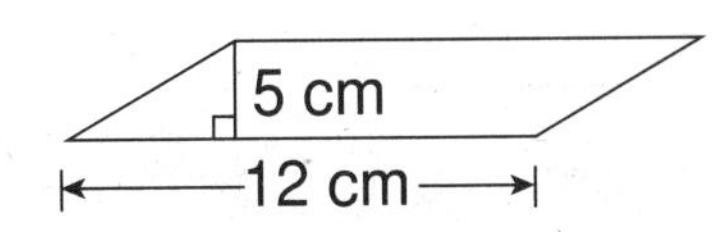

Area of parallelogram *WXYZ*
= $b \times h$

= ______ × ______ = ______ cm^2

8. In the rectangle graphed on the coordinate plane:

base = ______ units

height = ______ units.

Area of rectangle
= base × height

= ______ × ______

= ______ units^2

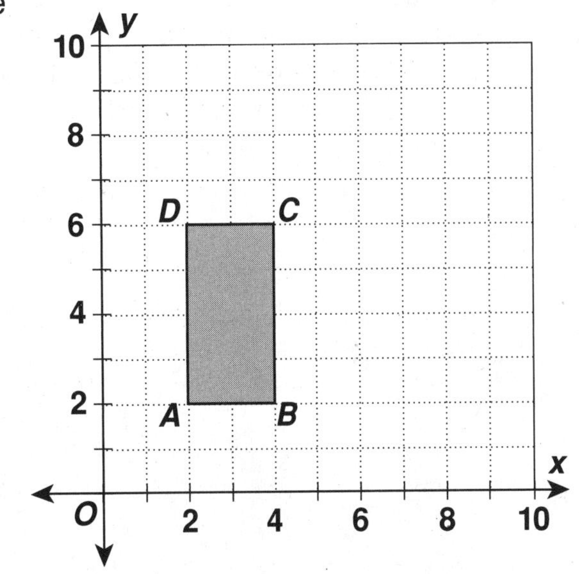

Name ______________________ Date __________ Class __________

LESSON **8-1**

Challenge
Color Me Least!

The basic rule for coloring a map is that no two regions that share a boundary can be the same color. However, two regions that meet at only a single point may have the same color.

In 1852, while coloring a map of England, Francis Guthrie noticed that no more than 4 colors were necessary. He conjectured that any map could be colored with no more than four colors.

What came to be known as the **Four Color Map Problem** was considered by mathematicians and school children alike for many years. No satisfactory proof was found until 1976, when K. Appel and W. Haken of the University of Illinois devised a computer program that took 1200 hours to run.

1. a. In this map of 5 distinct regions, can regions *C* and *D* have the same color? Explain.

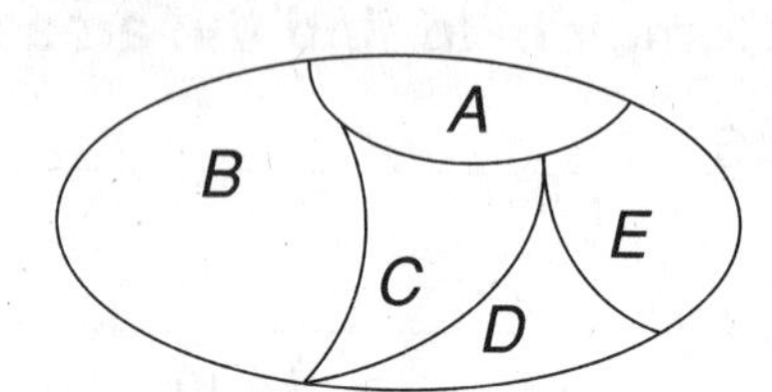

____

b. Can regions *C* and *E* have the same color? Explain.

____

c. What is the least number of colors required for this map? ______
Use numbers to show your answer on the map.

2. Here is a map of 10 neighboring states.
So far, as colored, only 3 colors are needed to distinguish among 9 of the 10.
Color 1: New Mexico, Nevada, and Wyoming
Color 2: Oregon, Arizona, and Montana
Color 3: Idaho, Colorado, and California
How can you color the state of Utah?

____

3. Use numbers to color these maps with the least number of colors possible.

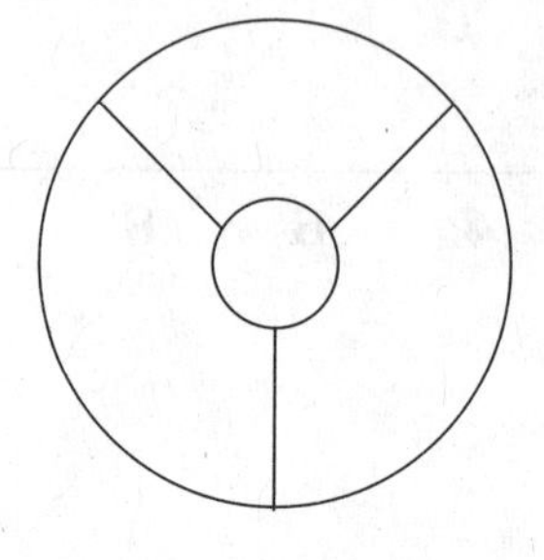

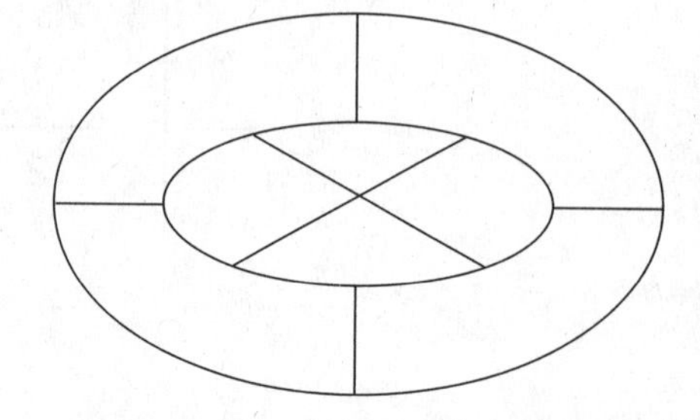

Name ______________________ Date __________ Class __________

LESSON 8-1

Problem Solving

Perimeter and Area of Rectangles and Parallelograms

Use the following for Exercises 1–2. A quilt for a twin bed is 68 in. by 90 in.

1. What is the area of the backing applied to the quilt?

   ____

2. A ruffle is sewn to the edge of the quilt. How many feet of ruffle are needed to go all the way around the edge of the quilt?

   ____

Use the following for Exercises 3–4. Jaime is building a rectangular dog run that is 12 ft by 8 ft.

3. If the run is cemented, how many square feet will be covered by cement?

   ____

4. How much fencing will be required to enclose the dog run?

   ____

Use the following for Exercises 5–6. Jackie is painting the walls in a room. Two walls are 12 ft by 8 ft, and two walls are 10 ft by 8 ft. Choose the letter for the best answer.

5. What is the area of the walls to be painted?

   A 352 ft^2 **C** 704 ft^2
   B 176 ft^2 **D** 400 ft^2

6. If a can of paint covers 300 square feet, how many cans of paint should Jackie buy?

   F 1 **H** 3
   G 2 **J** 4

Use the following for Exercises 7–8. One kind of pool cover is a tarp that stretches over the area of the pool and is tied down on the edge of the pool. The cover extends 6 inches beyond the edge of the pool. Choose the letter for the best answer.

7. A rectangular pool is 20 ft by 10 ft. What is the area of the tarp that will cover the pool?

   A 200 ft^2 **C** 60 ft^2
   B 231 ft^2 **D** 215.25 ft^2

8. If the tarp costs $2.50 per square foot, how much will the tarp cost?

   F $500.00 **H** $150.00
   G $538.13 **J** $577.50

Name ______________________ Date ____________ Class ____________

LESSON **8-1**

Reading Strategies

Understanding Symbols in a Formula

Perimeter is the distance around a figure.
The distance around rectangle *A* is
5 inches + 5 inches + 8 inches + 8 inches = 26 inches.

8 in.

5 in. *A*

You can find the **p**erimeter of a rectangle by adding 2 times the Base plus 2 times the Height.

Symbols are used to show this. ➤ $P = (2 \cdot b) + (2 \cdot h)$

The symbols make up the **formula** for finding the perimeter of a rectangle.

1. In the formula $P = (2 \cdot b) + (2 \cdot h)$, what does P stand for?

____

2. In the formula $P = (2 \cdot b) + (2 \cdot h)$, what does $(2 \cdot h)$ stand for?

____

Area is the amount of surface a figure covers. Area is measured in square units. You can count the square inches in rectangle *C* ➤ 20 square inches.

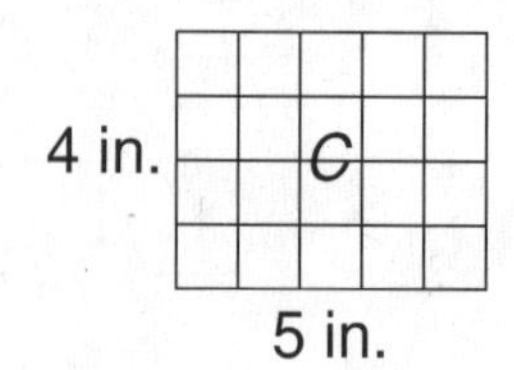

5 in.

You can find the **a**rea of a rectangle by multiplying the **b**ase times the **h**eight. Symbols are used to show this ➤ $A = b \cdot h$.

The symbols make up the formula for finding the area of a rectangle.

3. In the formula $A = b \cdot h$, what does A stand for? ____________________

4. In the formula $A = b \cdot h$, what does b stand for? ____________________

Name ______________________ Date ___________ Class ___________

LESSON **8-1**

Puzzles, Twisters & Teasers

What Floats Your Boat?

Find the perimeter of each figure. Each answer has a corresponding letter. Use the letters to solve the riddle.

1. $p =$ ______ **A**

9 m

11 m

2. $p =$ ______ **L**

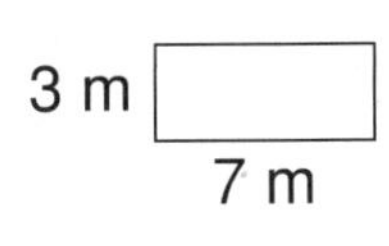

3. $p =$ ______ **E**

2 m
2 m
5 m
3 m
5 m
4 m
3 m
3 m
6 m

4. $p =$ ______ **I**

7 m

4 m

5. $p =$ ______ **M**

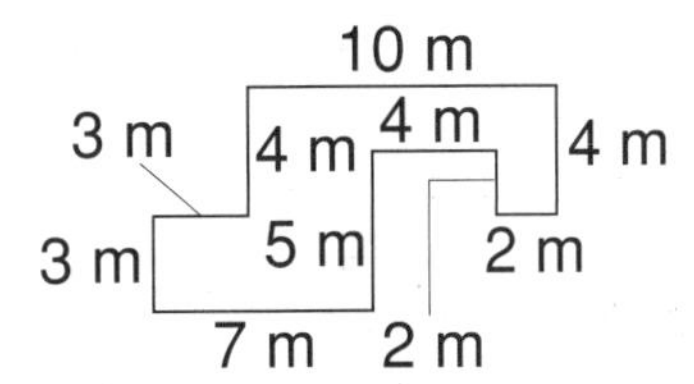

6. $p =$ ______ **R**

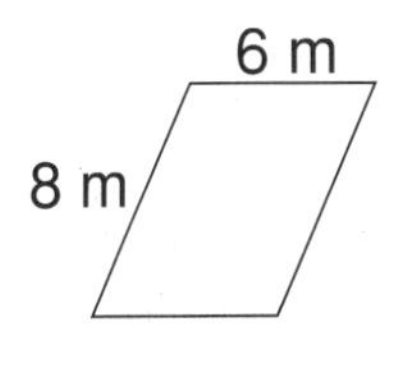

You were walking on a bridge and you saw a boat, yet there was not a single person on it. Why?

They were ____ L ____ ____ ____ R ____ ____ ____ D.

____	L	____	____	____	R	____	____	____	D.
40		20	44	40		28	22	36	

Name ______________________ Date __________ Class __________

LESSON 8-2

Practice A

Perimeter and Area of Triangles and Trapezoids

Find the perimeter of each figure.

1.

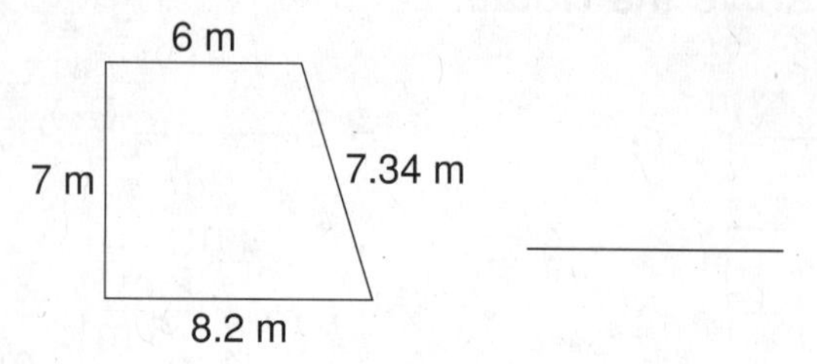

____

2.

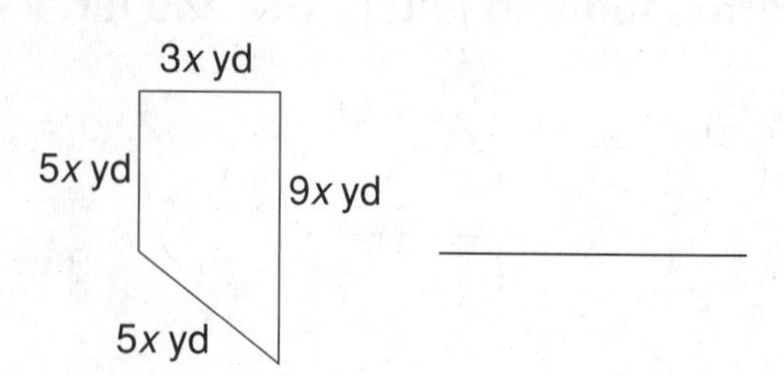

____

Find the missing measurement for each figure with the given perimeter.

3. triangle with perimeter 21 units

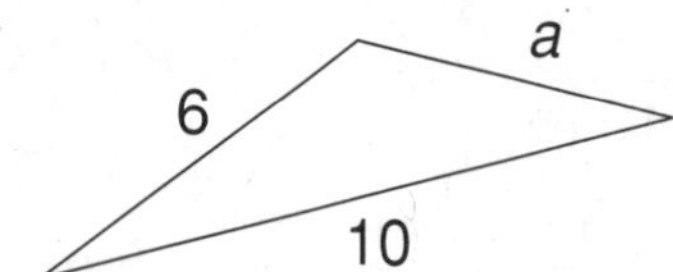

____

4. trapezoid with perimeter 39 units

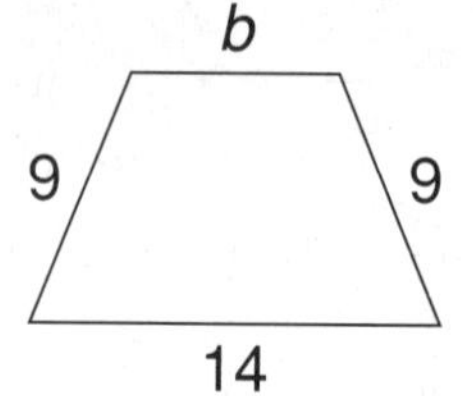

____

Graph and find the area of the figure with the given vertices.

5. $(-1, 4)$, $(-4, -3)$, $(4, -3)$

____

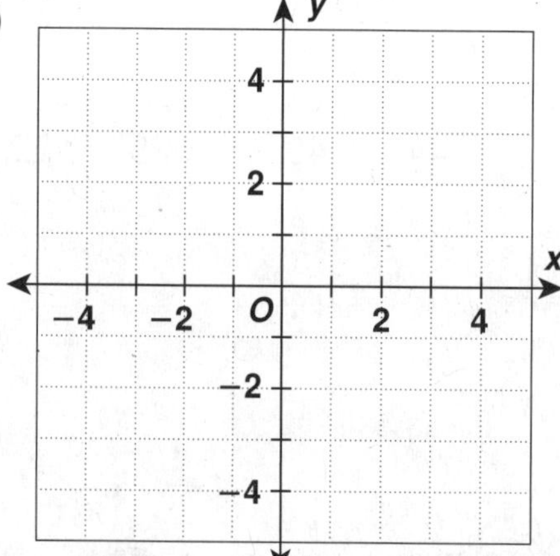

6. A garden shaped like a right triangle has two sides that measure 15 meters and 20 meters. Susie wants to put a fence along the perimeter of the garden.

a. How long is the third side of the garden? __________

b. How many meters of fencing material does she need? __________

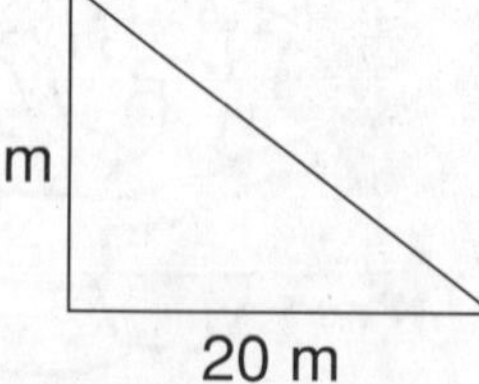

Name ______________________ Date ____________ Class ____________

LESSON 8-2

Practice B

Perimeter and Area of Triangles and Trapezoids

Find the perimeter of each figure.

1.

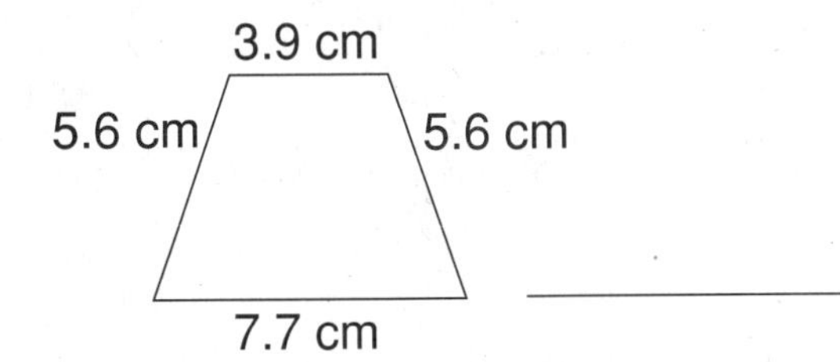

____

2.

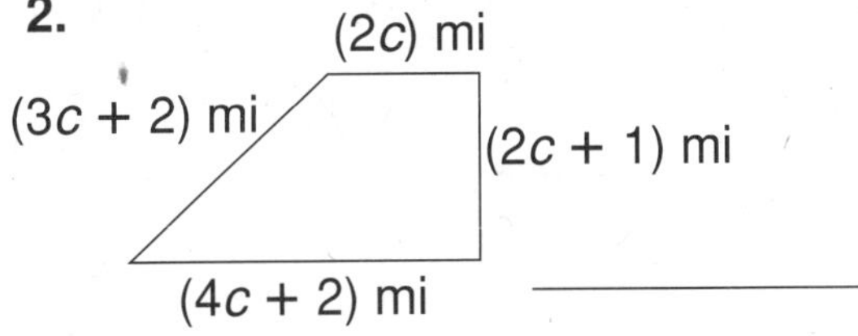

____

Find the missing measurement for each figure with the given perimeter.

3. triangle with perimeter 54 units

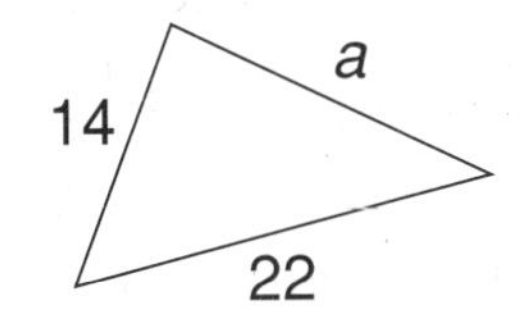

____

4. trapezoid with perimeter 34 units

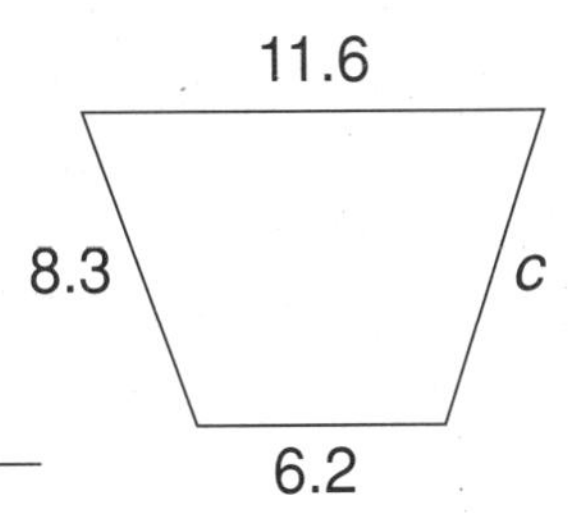

____

Graph and find the area of each figure with the given vertices.

5. $(-1, 3), (4, 3), (4, -4), (-4, -4)$

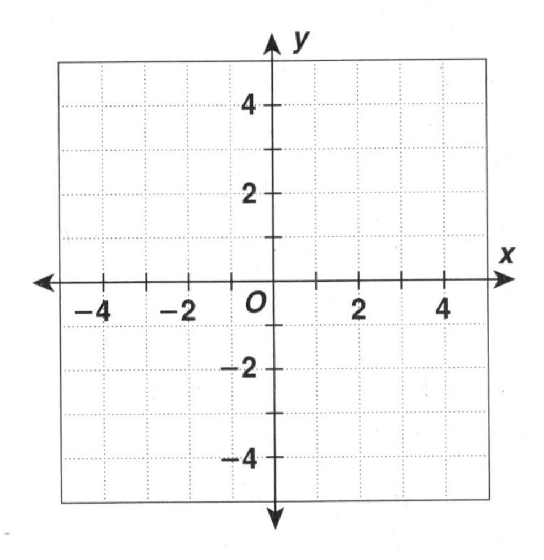

____

6. $(-1, 2), (-4, -2), (4, -2)$

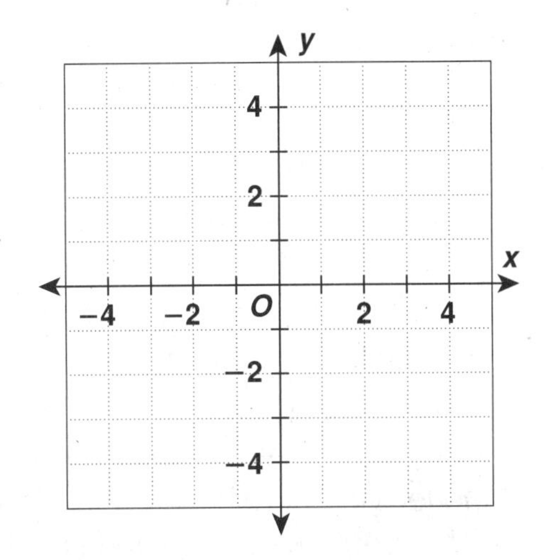

____

7. The two shortest sides of a pennant shaped like a right triangle measure 10 inches and 24 inches. Hank wants to put colored tape around the edge of the pennant. How many inches of tape does he need?

____

Name ______________________ Date ____________ Class ____________

LESSON **8-2**

Practice C

Perimeter and Area of Triangles and Trapezoids

Graph and find the area of the figure with the given vertices.

1. (−1, 2), (−1, −2), (−3, −4), (−3, 4)

____

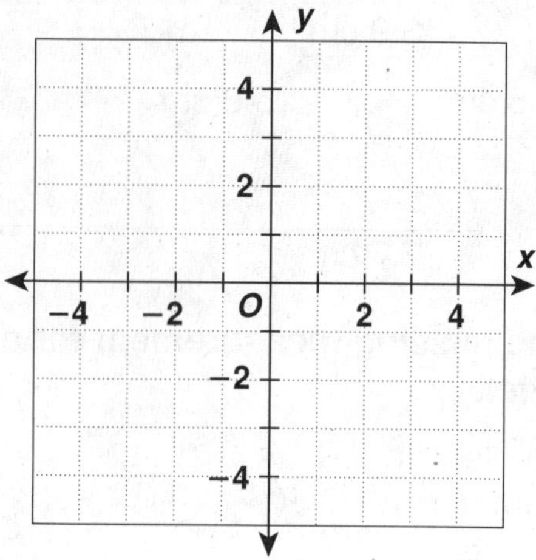

Find the area of each figure with the given dimensions.

2. triangle: $b = 10\frac{2}{3}$, $h = 12\frac{5}{6}$

____

3. trapezoid: $b_1 = 15.2$, $b_2 = 7.16$, $h = 14.5$

____

4. trapezoid: $b_1 = 9\frac{7}{8}$, $b_2 = 5\frac{3}{4}$, $h = 6\frac{2}{3}$

____

5. triangle: $b = 20.5$, $h = 17.64$

____

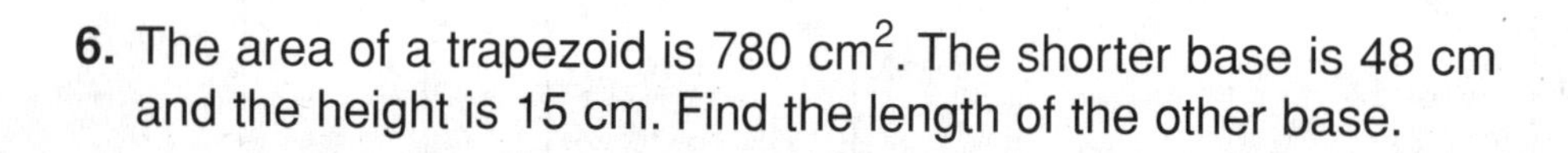

6. The area of a trapezoid is 780 cm^2. The shorter base is 48 cm and the height is 15 cm. Find the length of the other base.

____

7. Find the height of a triangle with area $85\frac{1}{3}$ ft^2 and base $10\frac{2}{3}$ ft.

____

8. Find the length of each side of a triangle if the perimeter is 73 in. The length of the second side is twice the length of the first side and the length of the third side is five more than the length of the first side.

____

9. The two shortest sides of a window shaped like a right triangle measure 24 centimeters and 32 centimeters. Bonnie wants to put a wooden frame around its edges. How many centimeters of wood are needed for the frame?

____

Name ______________________________ Date __________ Class __________

LESSON 8-2

Reteach

Perimeter and Area of Triangles and Trapezoids

To find the perimeter of a figure, add the lengths of all its sides.

Complete to find the perimeter of each figure.

1.

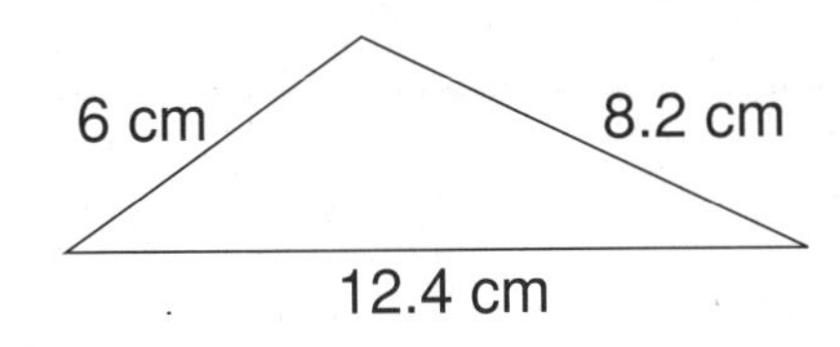

Perimeter of triangle

= _____ + _____ + _____

= _____ cm

2.

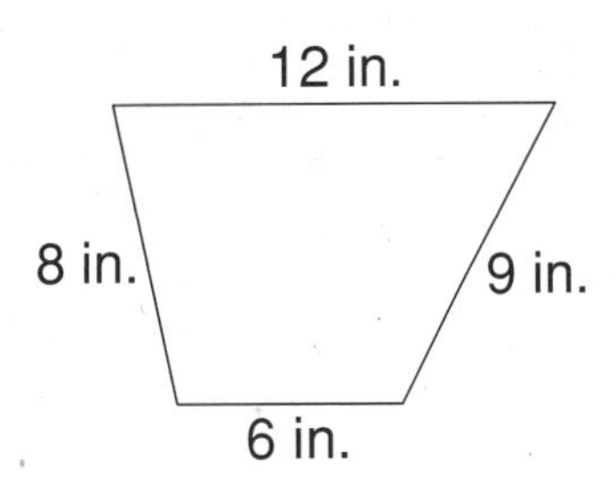

Perimeter of trapezoid

= _____ + _____ + _____ + _____

= _____ in.

Area of Triangle $= \frac{1}{2}bh$

The area of a triangle is one-half the product of a base length b and the height h drawn to that base.

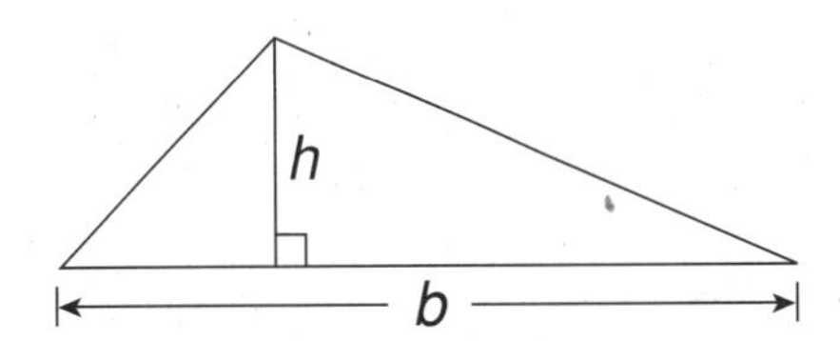

Complete to find the area of each triangle.

3. Area of triangle

$= \frac{1}{2}bh$

$= \frac{1}{2} \times$ _____ $\times$ _____

$= \frac{1}{2} \times$ _____ $=$ _____ in^2

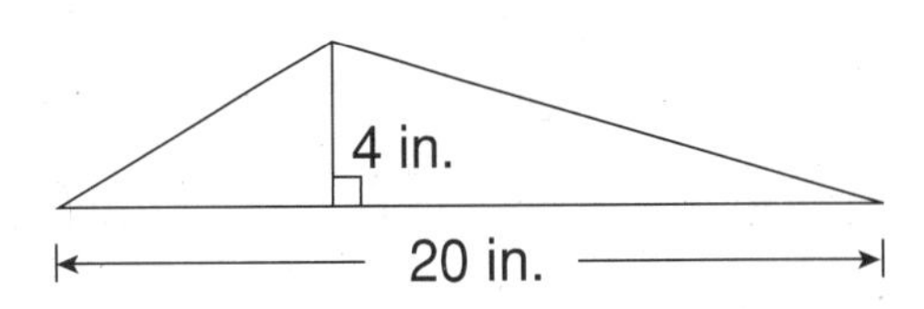

4. In the triangle graphed on the coordinate plane:

base = 10 − 3 = _____ units

height = 4 − (−2) = _____ units.

Area of triangle

$= \frac{1}{2} \times$ base $\times$ height

$= \frac{1}{2} \times$ _____ $\times$ _____

$= \frac{1}{2} \times$ _____ $=$ _____ units^2

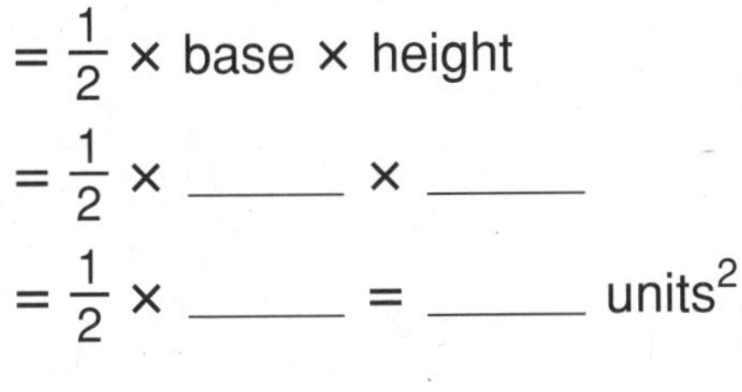

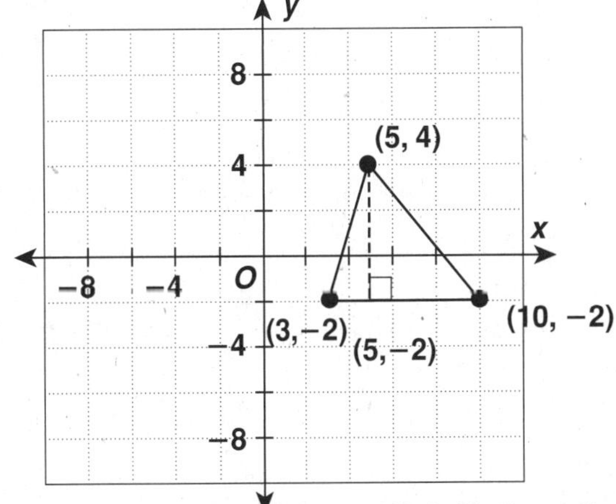

Name ______________________ Date __________ Class __________

LESSON 8-2

Reteach

Perimeter and Area of Triangles and Trapezoids (continued)

Area of Trapezoid $= \frac{1}{2}h(b_1 + b_2)$

The area of a trapezoid is one-half the height h times the sum of the base lengths b_1 and b_2.

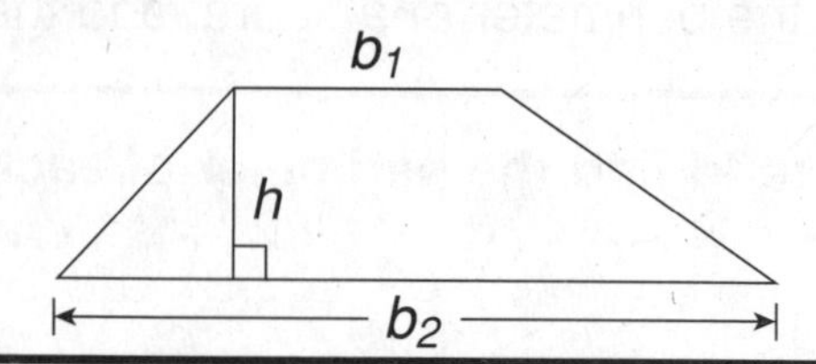

Complete to find the area of each trapezoid.

5.

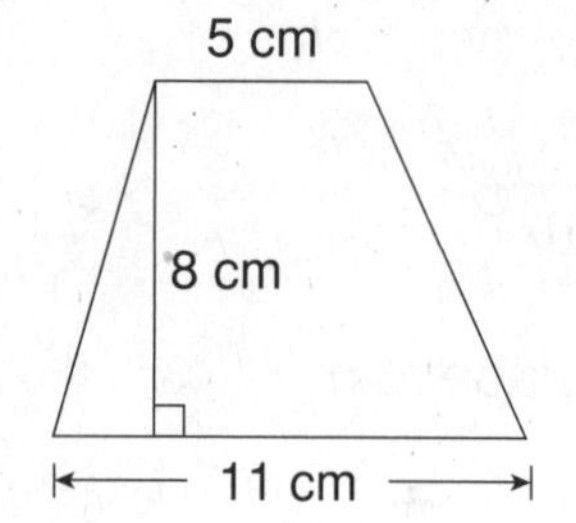

Area of trapezoid

$= \frac{1}{2}h(b_1 + b_2)$

$= \frac{1}{2} \times$ _____ $\times$ (_____ + _____)

$=$ _____ $\times$ (_____) $=$ _____ cm^2

6.

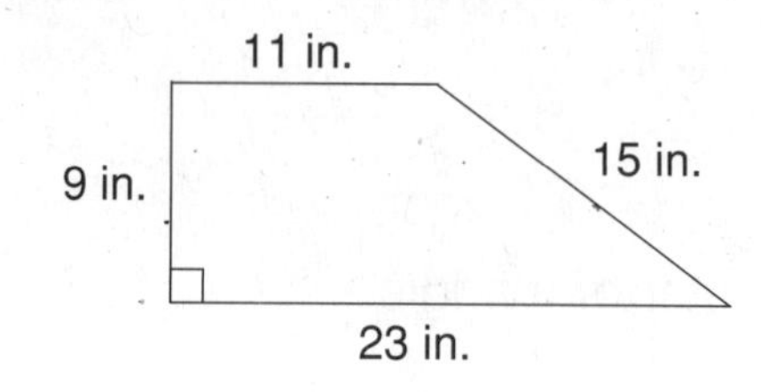

Area of trapezoid

$= \frac{1}{2}h(b_1 + b_2)$

$= \frac{1}{2} \times$ _____ $\times$ (_____ + _____)

$=$ _____ $\times$ (_____) $=$ _____ in^2

7. In the trapezoid graphed on coordinate plane:

$base_1 = 8 - 4 =$ _____ units

$base_2 = 11 - 2 =$ _____ units

height $= 6 - 2 =$ _____ units.

Area of trapezoid

$= \frac{1}{2} \times$ height $\times$ $(base_1 + base_2)$

$= \frac{1}{2} \times$ _____ $\times$ (_____ + _____)

$=$ _____ $\times$ (_____)

$=$ _____ $units^2$

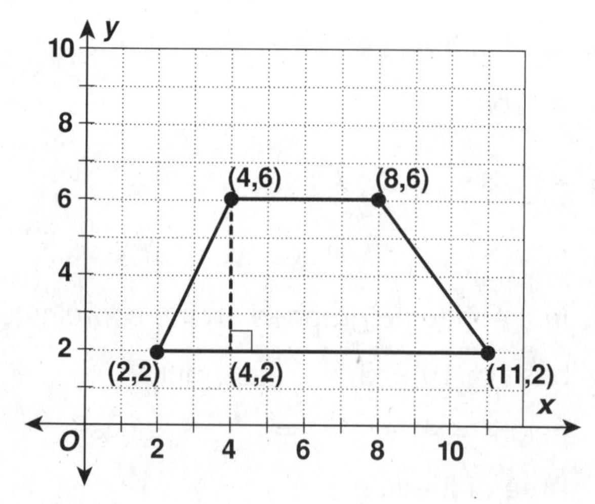

Name ______________________ Date ____________ Class ____________

LESSON 8-2

Challenge

Fence Me In!

You can find the area of a triangle in the coordinate plane that has no horizontal or vertical side.

Consider $\triangle ABC$ with vertices $A(-3, 2)$, $B(8, -3)$, and $C(5, 6)$.

By drawing horizontal and vertical lines, $\triangle ABC$ is enclosed in rectangle $PQBR$.

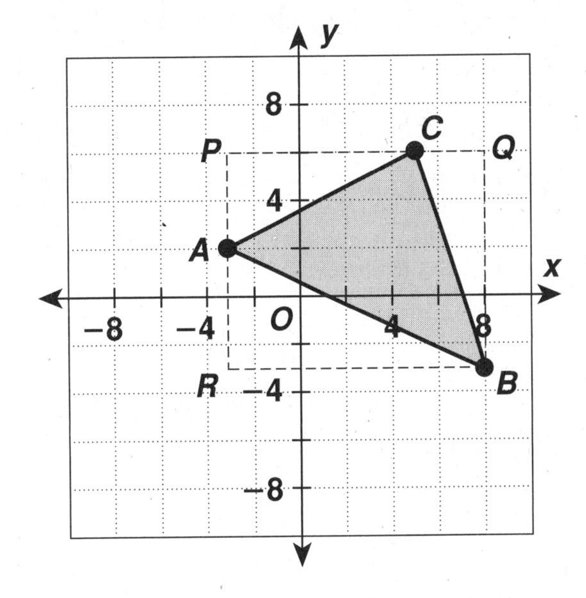

Write the coordinates of the remaining vertices of the rectangle.

1. *P* ____________

Q ____________

R ____________

Count boxes or subtract coordinates to find the indicated dimensions. Then find the indicated areas.

2. For rectangle *PQBR*:

base RB = ______ units height RP = ______ units area = ______ units^2

3. For right triangle *APC*:

base PC = ______ units height AP = ______ units area = ______ units^2

4. For right triangle *CQB*:

base CQ = ______ units height BQ = ______ units area = ______ units^2

5. For right triangle *ARB*:

base RB = ______ units height AR = ______ units area = ______ units^2

6. Explain how to combine the areas of the rectangle and the three right triangles to find the area of $\triangle ABC$. Then find the area of $\triangle ABC$.

____

____

Name ______________________ Date __________ Class __________

LESSON 8-2

Problem Solving

Perimeter and Area of Triangles and Trapezoids

Write the correct answer.

1. Find the area of the material required to cover the kite pictured below.

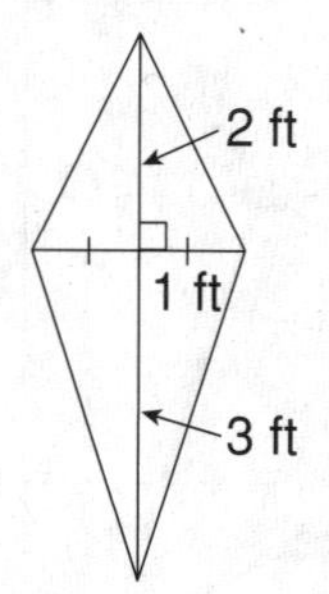

____

2. Find the area of the material required to cover the kite pictured below.

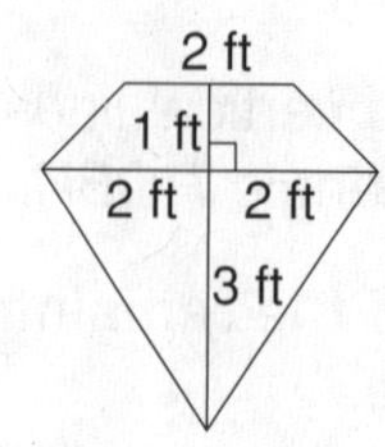

____

3. Find the approximate area of the state of Nevada.

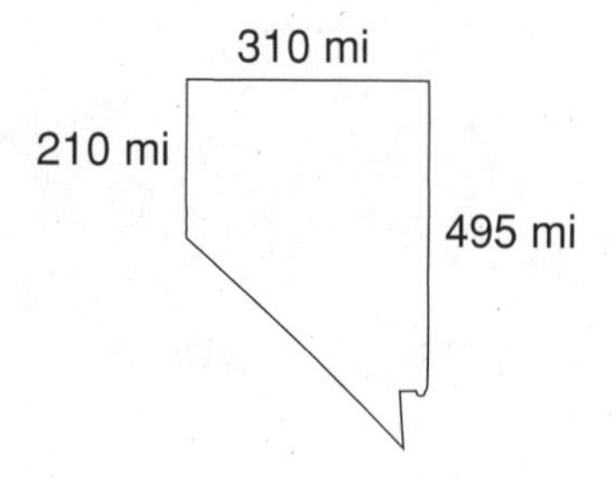

____

4. Find the area of the hexagonal gazebo floor.

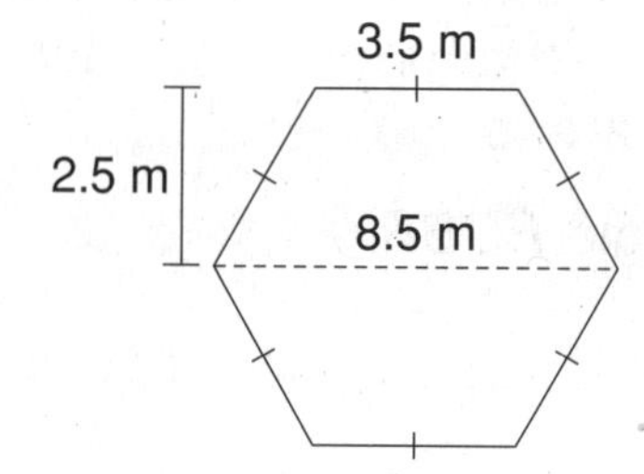

____

Choose the letter for the best answer.

5. Find the amount of flooring needed to cover the stage pictured below.

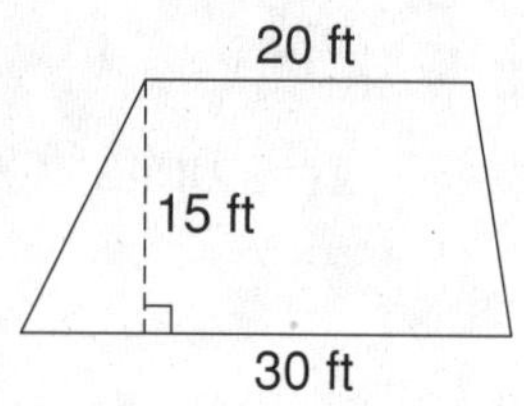

A 4500 ft^2

B 750 ft^2

C 525 ft^2

D 375 ft^2

6. Find the combined area of the congruent triangular gables.

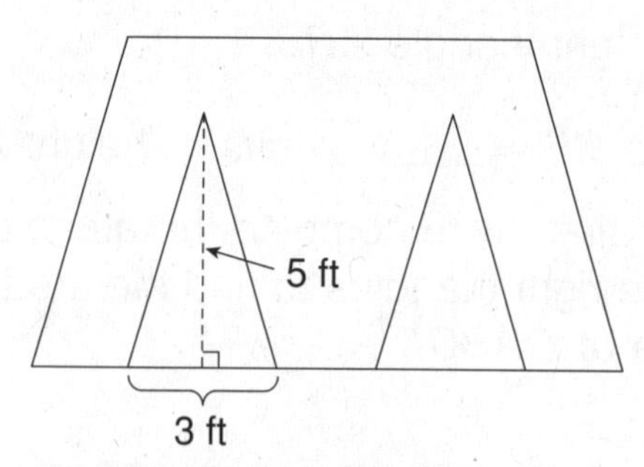

F 7.5 ft^2

G 15 ft^2

J 60 ft^2

H 30 ft^2

Name ____________________ Date ____________ Class ____________

LESSON 8-2

Reading Strategies

Compare and Contrast

Compare methods for finding the perimeter of rectangles and triangles.

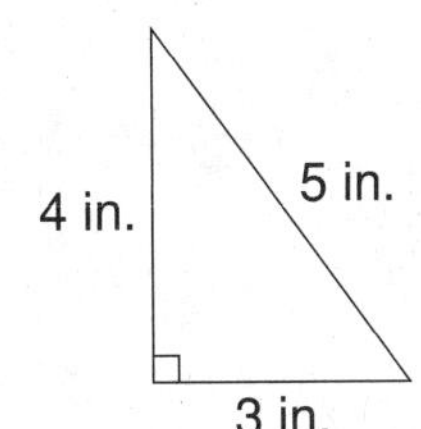

2 in.

8 in.

Perimeter of Triangles
Add lengths of all 3 sides.
$P = a + b + c$

Perimeter of Rectangles
Add lengths of all 4 sides.
$P = b + h + b + h$

Answer the questions to compare finding the perimeter of rectangles and triangles.

1. How is finding the perimeter of triangles and rectangles alike?

____

2. How is finding the perimeters of these figures different?

____

Compare methods for finding the area of rectangles and triangles.

Area of Rectangles

$A = \text{base} \cdot \text{height}$

$A = b \cdot h$

height 4 in.

3 in.
base

Area of Triangles

$A = \frac{1}{2} \cdot \text{base} \cdot \text{height}$

$A = \frac{1}{2} bh$

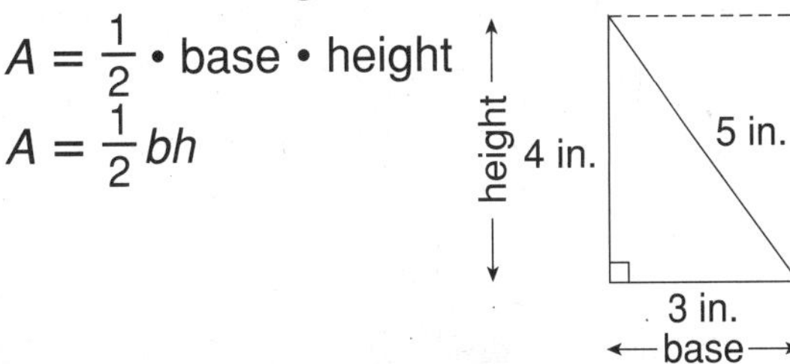

Answer the questions to compare finding the area of rectangles and triangles.

3. How does the area of this triangle compare to the area of this rectangle?

____

4. How is finding the area of this triangle different than finding the area of this rectangle?

____

____

Name ______________________ Date __________ Class __________

LESSON 8-2

Puzzles, Twisters & Teasers

A Whole Hole!

Find the perimeter of each figure. Each answer has a corresponding letter. Use the letters to solve the riddle.

1. $p =$ ______ **E**

8 ft, 10 ft, 14 ft

2. $p =$ ______ **M**

7 ft, 2 ft, 4 ft, 11 ft

3. $p =$ ______ **A**

4 ft, 4 ft, 4 ft

4. $p =$ ______ **H**

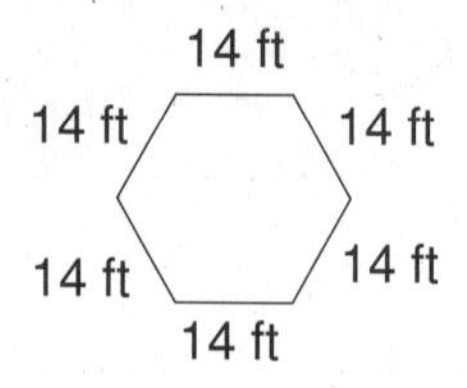

5. $p =$ ______ **K**

12 ft, 12 ft, 6 ft

6. $p =$ ______ **O**

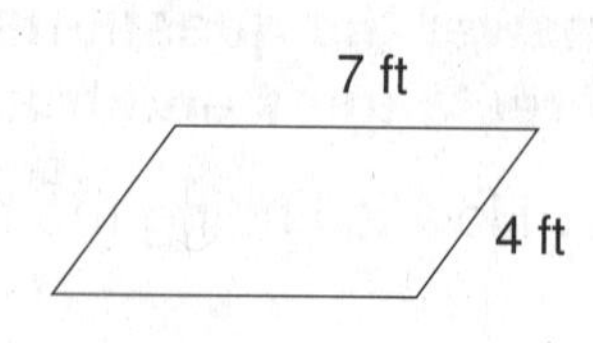

Suppose you were in a room with no door or windows, but two halves of a table. How could you get out?

Put the halves together to ___ ___ ___ E A W ___ ___ L ___

___	___	___	E	A	W	___	___	L	___
24	12	30				84	22		32

Name ______________________ Date __________ Class __________

LESSON 8-3

Practice A
Circles

Find the circumference of each circle, both in terms of π and to the nearest tenth. $C = \pi d$. Use 3.14 for π.

1. circle with diameter 4 cm

____

2. circle with radius 4 in.

____

3. circle with radius 5.5 ft

____

4. circle with diameter 3 m

____

5. circle with radius 9 cm

____

6. circle with diameter 8.2 in.

____

Find the area of each circle, both in terms of π and to the nearest tenth. $A = \pi r^2$. Use 3.14 for π.

7. circle with radius 5 m

____

8. circle with diameter 8 ft

____

9. circle with diameter 6 cm

____

10. circle with radius 9 in.

____

11. circle with radius 3.1 m

____

12. circle with diameter 4.8 cm

____

13. Graph a circle with center (0, 0) that passes through (0, 2). Find the area and circumference, both in terms of π and to the nearest tenth. Use 3.14 for π.

____

____

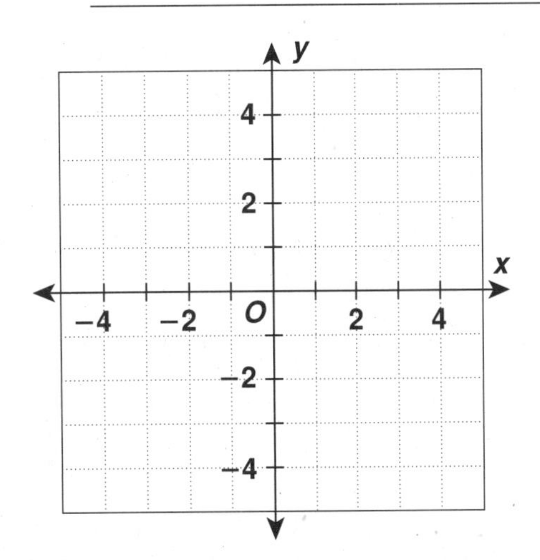

14. A wheel has a diameter of 21 inches. About how far does it travel if it makes 30 complete revolutions? Use $\frac{22}{7}$ for π.

____

Name ______________________ Date ____________ Class ____________

LESSON **8-3**

Practice B
Circles

Find the circumference of each circle, both in terms of π and to the nearest tenth. Use 3.14 for π.

1. circle with radius 10 in.

____

2. circle with diameter 13 cm

____

3. circle with diameter 18 m

____

4. circle with radius 15 ft

____

5. circle with radius 11.5 in.

____

6. circle with diameter 16.4 cm

____

Find the area of each circle, both in terms of π and to the nearest tenth. Use 3.14 for π.

7. circle with radius 9 in.

____

8. circle with diameter 14 cm

____

9. circle with radius 20 ft

____

10. circle with diameter 17 m

____

11. circle with diameter 15.4 m

____

12. circle with radius 22 yd

____

13. Graph a circle with center (0, 0) that passes through (0, −3). Find the area and circumference, both in terms of π and to the nearest tenth. Use 3.14 for π.

____

____

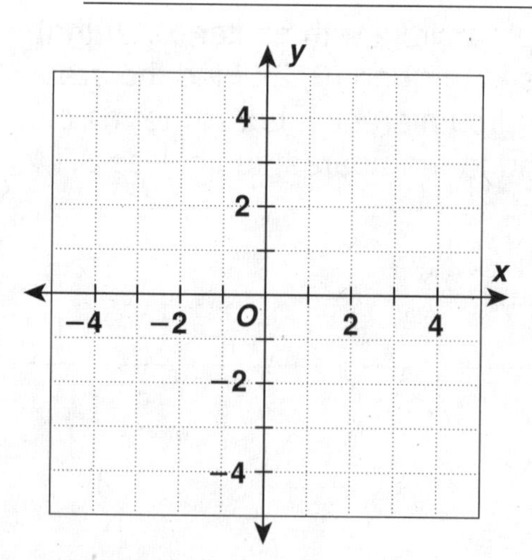

14. A wheel has a radius of $2\frac{1}{3}$ feet. About how far does it travel if it makes 60 complete revolutions? Use $\frac{22}{7}$ for π.

____

Name ________________________ Date __________ Class __________

LESSON **8-3**

Practice C
Circles

Find the circumference and area of each circle to the nearest tenth. Use 3.14 for π.

1.

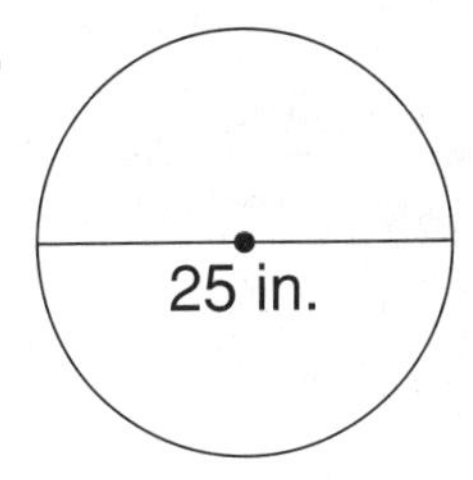

2.

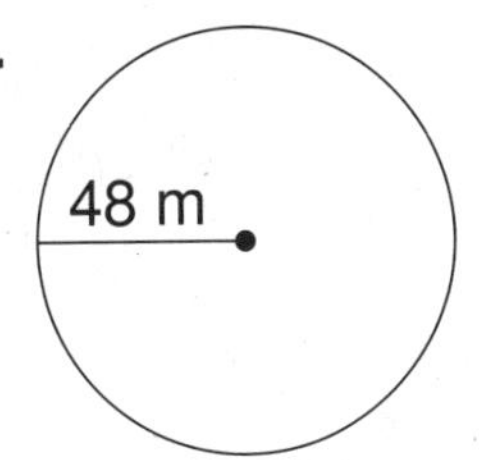

3.

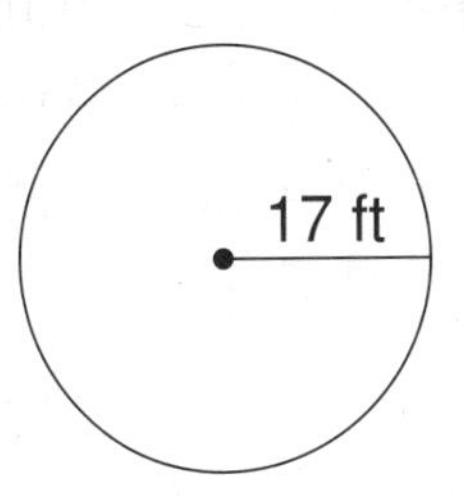

________________ ________________ ________________

________________ ________________ ________________

Find the radius of each circle with the given measurement.

4. $C = 36\pi$ cm ________

5. $C = 42\pi$ ft ________

6. $C = 80.1\pi$ m ________

7. $C = 152.6\pi$ in. ________

8. $A = 121\pi\ \text{cm}^2$ ________

9. $A = 734.41\pi\ \text{ft}^2$ ________

10. $A = 1024\pi\ \text{m}^2$ ________

11. $A = 5184\pi\ \text{in}^2$ ________

Find the shaded area to the nearest tenth. Use 3.14 for π.

12.

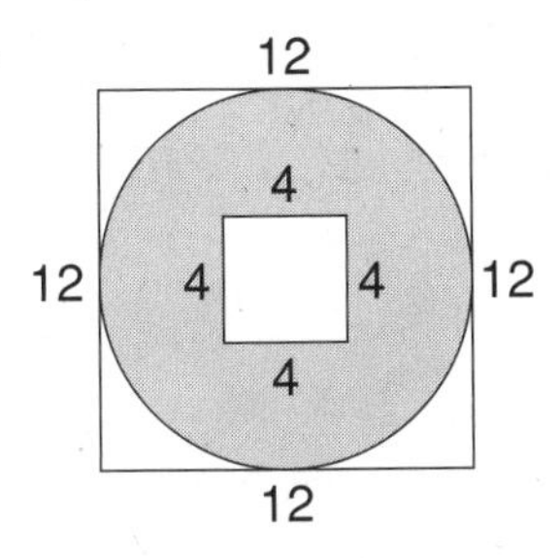

13.

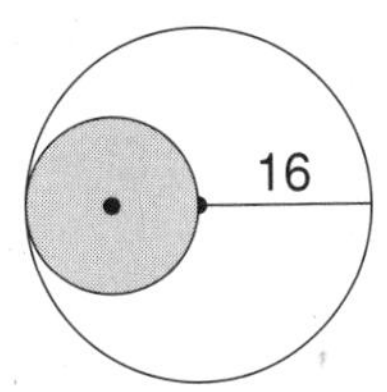

14.

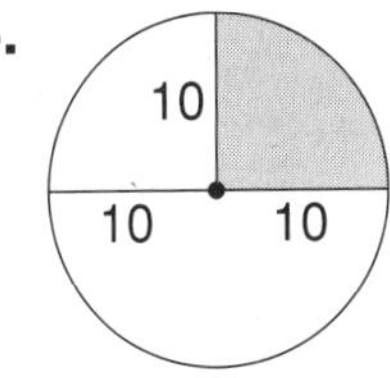

________________ ________________ ________________

15. Derek is riding his bicycle with a 36-in. diameter wheel in a 10-mi race. How many revolutions will each wheel make in the race? Round your answer to the nearest tenth. Use 3.14 for π. (Hint: 12 in. = 1 ft; 5280 ft = 1 mi)

____

Name ______________________________ Date __________ Class __________

CHAPTER 8-3

Reteach

Circles

A **radius** connects the **center** of a **circle** to any point on the circle.

A **diameter** passes through the center and connects two points on the circle.

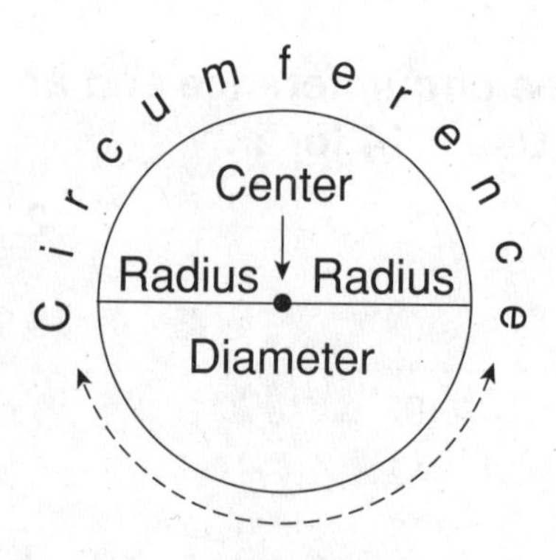

diameter d = twice radius r

$\mathbf{d = 2r}$

Circumference is the distance around a circle.

(The symbol ≈ means *is approximately equal to.*)

Circumference $C \approx 3(\text{diameter } d)$

$\mathbf{C = \pi d}$

For a circle with diameter = 8 in.

$C = \pi d$

$C = \pi(8)$

$C = 8\pi$ in.

$\pi \approx 3.14$ $\quad C \approx 8(3.14) \approx 25.12$ in.

Circumference $C \approx 6(\text{radius } r)$

$\mathbf{C = 2\pi r}$

For a circle with radius = 8 in.

$C = 2\pi r$

$C = 2\pi(8)$

$C = 16\pi$ in.

$\pi \approx 3.14$ $\quad C \approx 16(3.14) \approx 50.24$ in.

Find the circumference of each circle, exactly in terms of π and approximately when $\pi = 3.14$.

1. diameter = 15 ft

   $C = \pi d$

   $C = \pi($______$) =$ ______ ft

   $C \approx 3.14($______$) \approx$ ______ ft

2. radius = 4 m

   $C = 2\pi r$

   $C = 2\pi($______$) =$ ______ m

   $C \approx$ ______$(3.14) \approx$ ______ m

Area $A \approx 3(\text{the square of radius } r)$

$\mathbf{A = \pi r^2}$

For a circle with radius = 5 in.: $A = \pi r^2 = \pi(5^2) = 25\pi \text{ in}^2$

$A \approx 25(3.14) \approx 78.5 \text{ in}^2$

Find the area of each circle, exactly in terms of π and approximately when $\pi = 3.14$.

3. radius = 9 ft

   $A = \pi r^2$

   $A = \pi($______$) =$ ______ ft^2

   $A \approx$ ______$(3.14) \approx$ __________ ft^2

4. diameter = 10 m, radius = ______ m

   $A = \pi r^2$

   $A = \pi($______$) =$ ______ m^2

   $A \approx$ ______$(3.14) \approx$ ______ m^2

Name ______________________ Date ____________ Class ____________

Challenge
LESSON 8-3 Circles

Work in $\triangle PQR$ at the bottom of this page.

1. Use a ruler to find the midpoints of the three sides of the triangle. Label these midpoints M_1, M_2, M_3.
2. An **altitude** of a triangle is a segment drawn from a vertex perpendicular to the opposite side. Draw the altitude to each side of the triangle. Use F_1, F_2, F_3 to label the foot of each altitude (where the altitude meets the side of the triangle at right angles).
3. The point at which the altitudes meet is the **orthocenter** of the triangle. Label the orthocenter O. Locate the midpoints of $\overline{OP}$, $\overline{OQ}$, $\overline{OR}$, the segments that connect orthocenter O to each vertex. Labels these midpoints D_1, D_2, D_3.
4. If you have been accurate in your measurements, the nine points – M_1, M_2, M_3, F_1, F_2, F_3, D_1, D_2, D_3 – lie on a circle. Draw the **nine-point circle** for $\triangle PQR$.

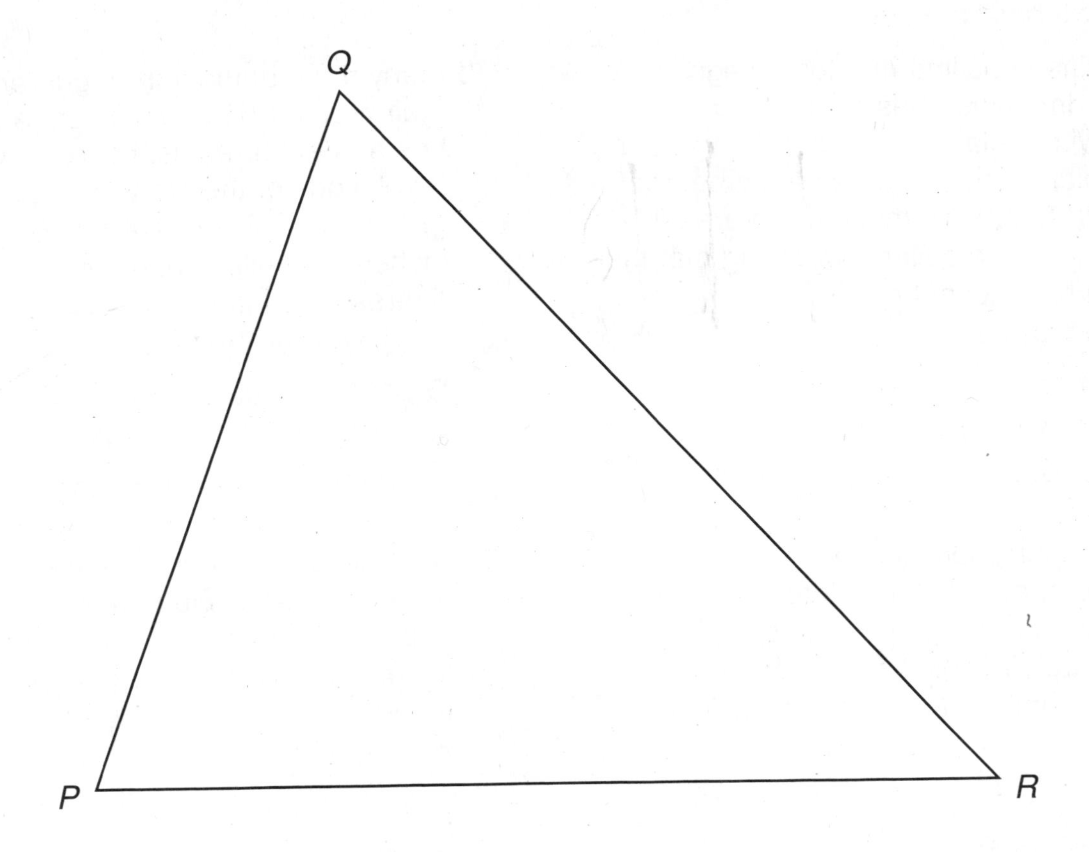

Name ______________________ Date ____________ Class ____________

LESSON **8-3**

Problem Solving

Circles

Round to the nearest tenth. Use 3.14 for π. Write the correct answer.

1. The world's tallest Ferris wheel is in Osaka, Japan, and stands 369 feet tall. Its wheel has a diameter of 328 feet. Find the circumference of the Ferris wheel.

   ____

2. A dog is on a 15-foot chain that is anchored to the ground. How much area can the dog cover while he is on the chain?

   ____

3. A small pizza has a diameter of 10 inches, and a medium has a diameter of 12 inches. How much more pizza do you get with the medium pizza?

   ____

4. How much more crust do you get with a medium pizza with a diameter of 12 inches than a small pizza with a 10 inch diameter?

   ____

Round to the nearest tenth. Use 3.14 for π. Choose the letter for the best answer.

5. The wrestling mat for college NCAA competition has a wrestling circle with a diameter of 32 feet, while a high school mat has a diameter of 28 feet. How much more area is there in a college wrestling mat than a high school mat?

   A 12.6 ft^2

   B 188.4 ft^2

   C 234.8 ft^2

   D 753.6 ft^2

6. Many tire manufacturers guarantee their tires for 50,000 miles. If a tire has a 16-inch radius, how many revolutions of the tire are guaranteed? There are 63,360 inches in a mile. Round to the nearest revolution.

   F 630.6 revolutions

   G 3125 revolutions

   H 31,528,662 revolutions

   J 500,000,000 revolutions

7. In men's Olympic discus throwing competition, an athlete throws a discus with a diameter of 8.625 inches. What is the circumference of the discus?

   A 13.5 in.

   B 27.1 in.

   C 58.4 in.

   D 233.6 in.

8. An athlete in a discus competition throws from a circle that is approximately 8.2 feet in diameter. What is the area of the discus throwing circle?

   F 52.8 ft^2

   G 25.7 ft^2

   H 12.9 ft^2

   J 211.1 ft^2

Name ________________ Date ________ Class ________

LESSON 8-3

Reading Strategies

Graphic Organizer

This chart will help you organize and understand key measurements for circles.

Circle Measurements

Radius

- A line segment from the center of the circle to a point on the circle

Diameter

- Twice the radius
- Line segment passes through center of circle
- Connects 2 points on the circle

Circumference

- The distance around a circle

Pi → π

- Ratio of circumference to diameter: → $\frac{C}{d}$
- Value of pi is about 3.14

Use the information in the chart to answer each question.

1. What is the distance around a circle called?

____

2. What is twice the length of the radius?

____

3. What is pi?

____

4. What is the estimated value of pi? ________________

5. What is the name of the line segment that extends from the center of the circle to a point on the circle?

____

The formula for finding the circumference of a circle is $C = \text{pi} \cdot d$.

6. If the length of the diameter is 6 feet, rewrite the formula using this value and the value for pi.

____

7. If you only knew the radius of a circle, could you still find the circumference? Why or why not?

____

Name ______________________ Date ____________ Class ____________

LESSON 8-3

Puzzles, Twisters & Teasers

Circular Reasoning!

Find the circumference of each circle to the nearest tenth. Each answer has a corresponding letter. Use the letters to solve the riddle.

1. $c =$ ________ **S**

3.3 cm

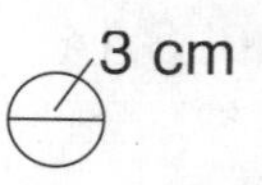

2. $c =$ ________ **W**

1 cm

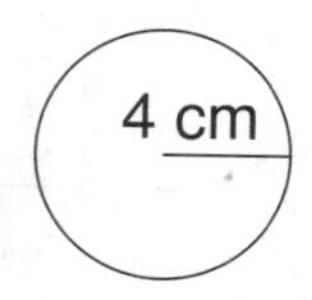

3. $c =$ ________ **H**

5 cm

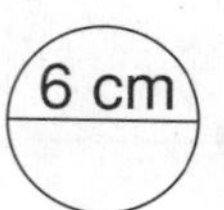

4. $c =$ ________ **E**

3 cm

5. $c =$ ________ **C**

4 cm

6. $c =$ ________ **N**

6 cm

How do you know when a stream needs oil?

___	___	E	___	IT	___	R	___	___	K	___.
3.1	15.7		18.8		25.1		9.4	9.4		20.7

Name ______________________ Date ____________ Class ____________

Practice A
Drawing Three-Dimensional Figures

1. Name the vertices, edges, and faces of the three-dimensional figure shown.

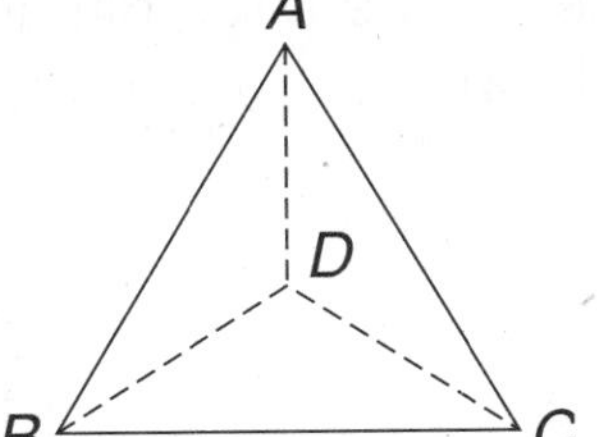

Vertices: ______________________

Edges: ______________________

Faces: ______________________

2. Draw the figure that has the following top, front, and side views.

Front　　Top　　Side

3. Draw the front, top, and side views of the figure.

Name ______________________ Date __________ Class __________

LESSON 8-4

Practice B

Drawing Three-Dimensional Figures

1. Name the vertices, edges, and faces of the three-dimensional figure shown.

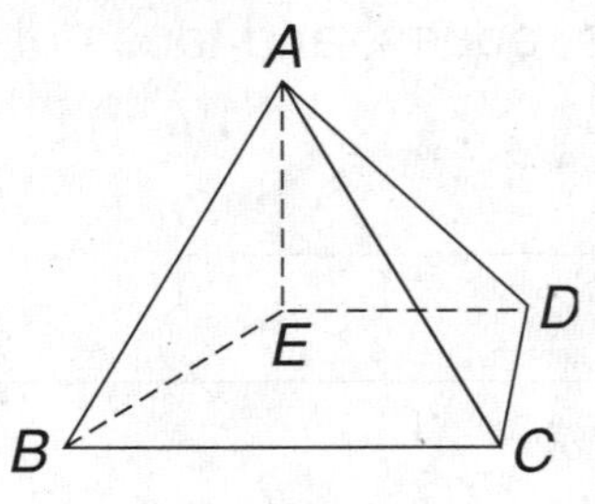

____

edges: ______________________________

faces: ______________________________________

2. Draw the figure that has the following top, front, and side views.

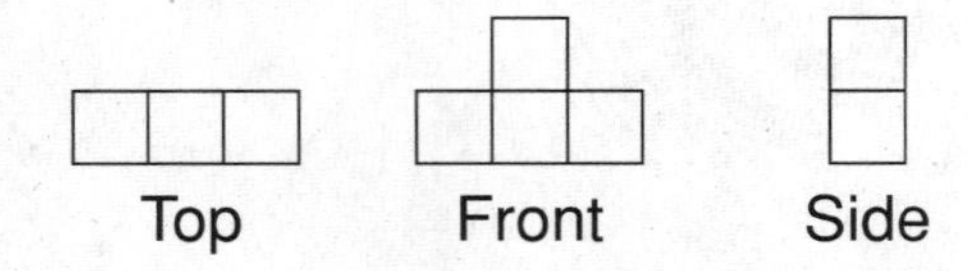

3. Draw the front, top, and side views of the figure.

Name ______________________ Date __________ Class __________

LESSON **8-4**

Practice C
Drawing Three-Dimensional Figures

Give the dimensions of the base and height in each figure. Name the faces that are parallel.

1. Draw the figure that has the following top, front, and side views.

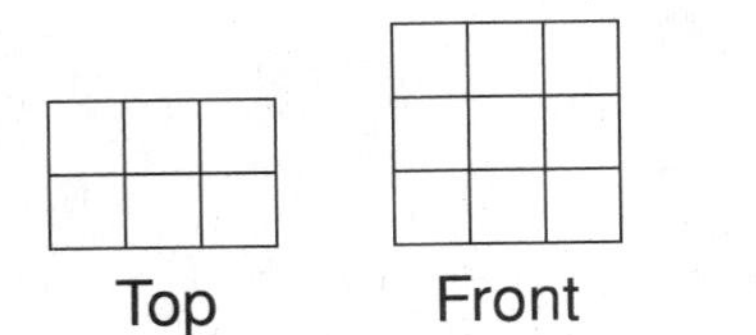

2. Draw the front, top, and side views of the figure.

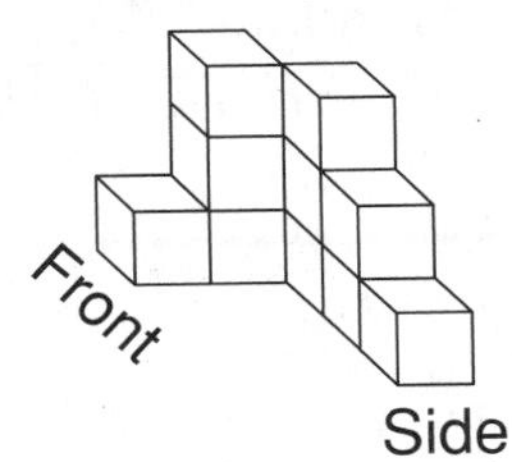

Use isometric dot paper to sketch each figure.

3. a triangular box 4 units high

4. a rectangular box 2 units high with a base 4 units by 2 units

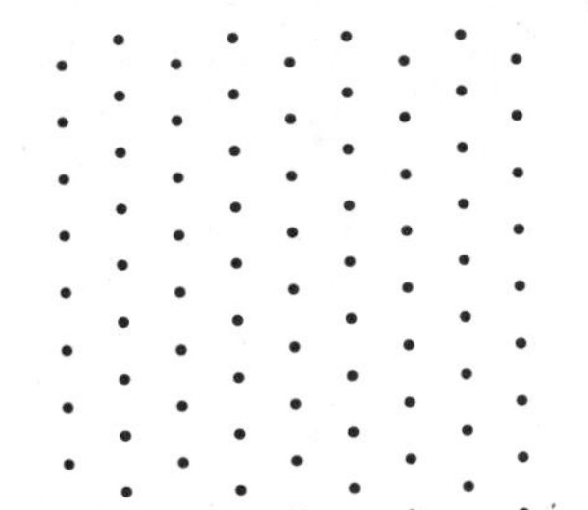

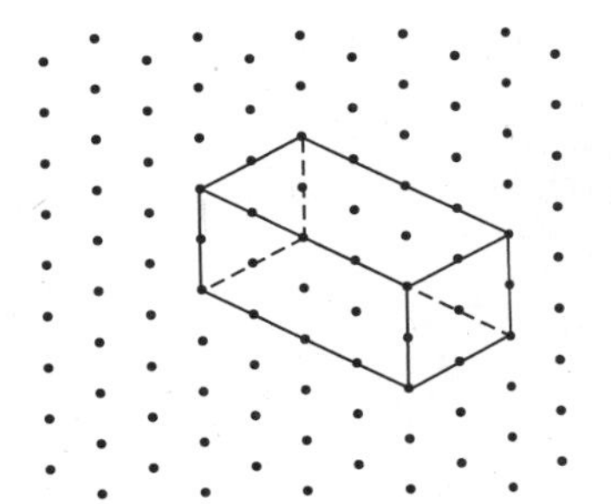

Name ______________________ Date __________ Class __________

LESSON 8-4

Reteach

Drawing Three-Dimensional Figures

Some elements of a three-dimensional figure are:

face, a flat surface that is a plane figure
edge, the line where two faces meet
vertex, the point where three or more edges meet. → Use **vertices** for more than one vertex.

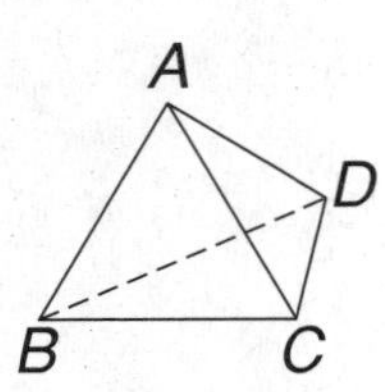

The triangular pyramid has 4 vertices: *A, B, C, D.*
The faces of the figure are triangles. The figure has 4 faces: triangles *ABC, ADC, ABD, BDC.*
The figure has 6 edges.
Trace the lines where the faces meet to name the edges: $\overline{AB}$, $\overline{AC}$, $\overline{AD}$, $\overline{BC}$, $\overline{BD}$, $\overline{CD}$.

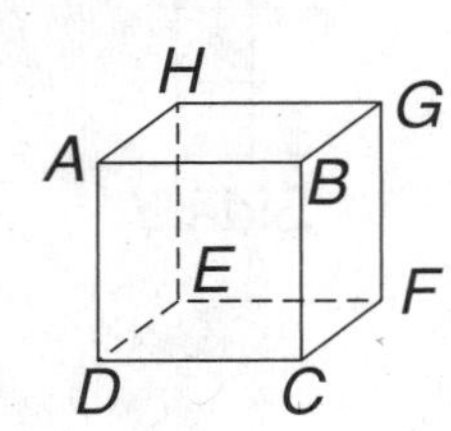

The cube has 8 vertices: *A, B, C, D, E, F, G, H.*
The faces of the figure are squares.
The cube has 6 faces: squares *ABCD, ABGH, AHED, EDCF, BCFG, EFGH.*
The cube has 12 edges: $\overline{AB}$, $\overline{BC}$, $\overline{CD}$, $\overline{AD}$, $\overline{AH}$, $\overline{BG}$, $\overline{CF}$, $\overline{DB}$, $\overline{EF}$, $\overline{FG}$, $\overline{GH}$, $\overline{HE}$.

Complete to name the vertices, faces, and edges of the figures shown.

1.

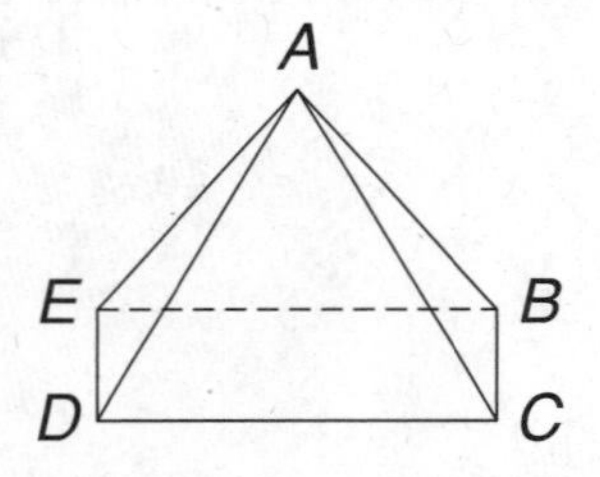

Vertices: *A, B,* _____, _____, _____

Faces: triangles *ABC, ACD,* _______, _______, rectangle _______

Edges: $\overline{AB}$, $\overline{AC}$, $\overline{AD}$, $\overline{AE}$, _______, _______, _______, _______

2.

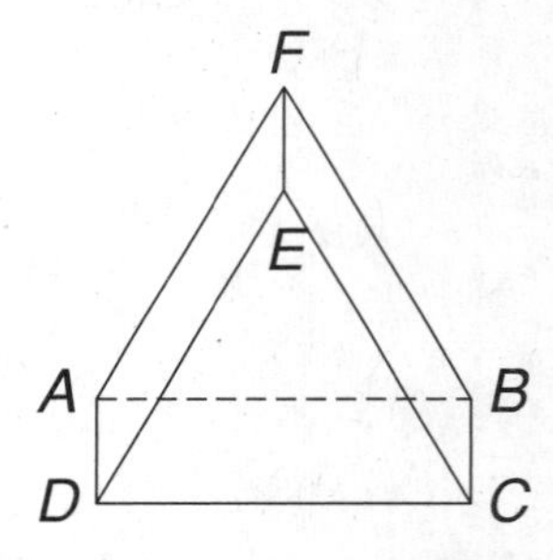

Vertices: *A, B,* _____, _____, _____, _____

Faces: triangles *ABF,* _____, rectangles ABCD, _______, _______

Edges: $\overline{AB}$, $\overline{BC}$, $\overline{CD}$, $\overline{AD}$, _____, _____, _____, _____, _____

Name ______________________ Date ____________ Class ____________

LESSON 8-4

Reteach

Drawing Three-Dimensional Figures (continued)

You can use the **front**, **side**, and **top** views of a three-dimensional figure to show how the figure looks from different perspectives. When the figures are constructed from cubes, the views will be different groups of squares.

From the front, the stack of cubes appears to have

1 cube on top,
2 cubes in the center, and
3 cubes on the bottom.

The figure looks like 1 square on top of 2 squares that are on top of 3 squares.

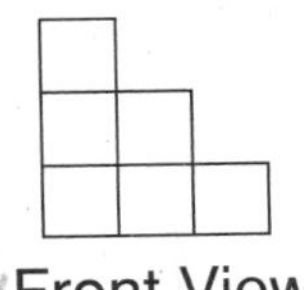

Front View

From the side, there appears to be a stack 3 cubes high.

The figure looks like 3 squares tall.

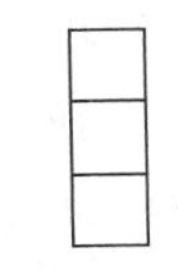

Side View

From the top, there appears to be a row 3 cubes long.

The figure looks like 3 squares long.

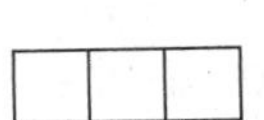

Top View

Complete to draw the front, side, and top view of the figure shown.

3.

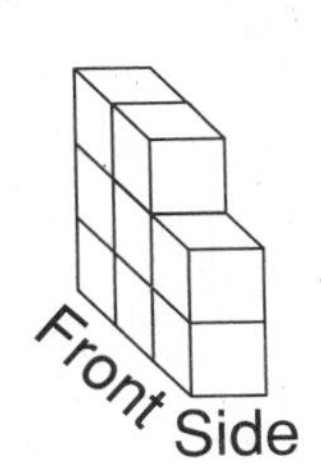

From the front, the stack of cubes appears to have _____ cubes on top, _____ cubes in the center, and _____ cubes on the bottom.

Front View

From the side, there appears to be a stack _____ cubes high.

Side View

From the top, there appears to be a row _____ cubes long.

The figure looks like 3 squares long.

Top View

Name ______________________ Date ____________ Class ____________

LESSON **8-4**

Challenge

From the Ground, Up!

Based on an *isometric drawing* of a solid figure, a **foundation drawing** shows only the base of the solid figure — as a pattern of squares, with a number in each square to tell how many cubes are above it.

Isometric Drawing

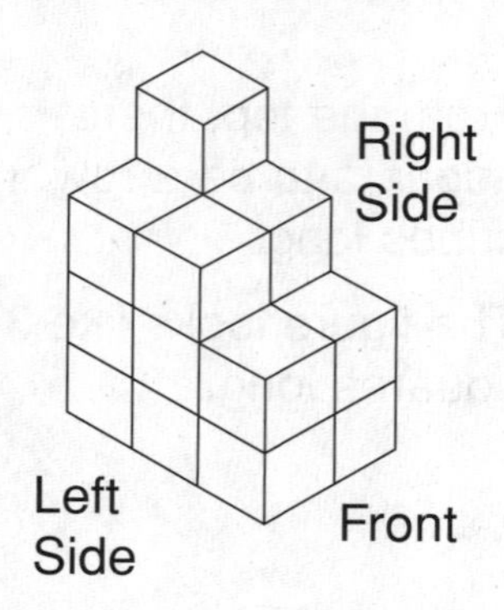

Foundation Drawing

Left Side / Right Side

3	4
3	3
2	2

Front

The front row of the figure is just two cubes high. So, there is a 2 in each box in the front row of the foundation drawing.

The right rear corner column, the highest in the figure, is four cubes high. So, there is a corresponding 4 in the foundation drawing.

Make a foundation drawing for each solid figure.

1.

2.

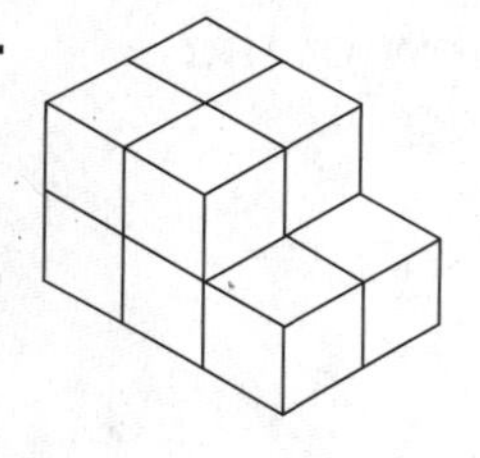

3.

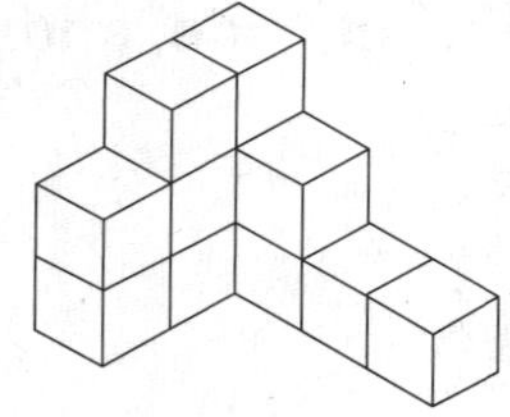

4.

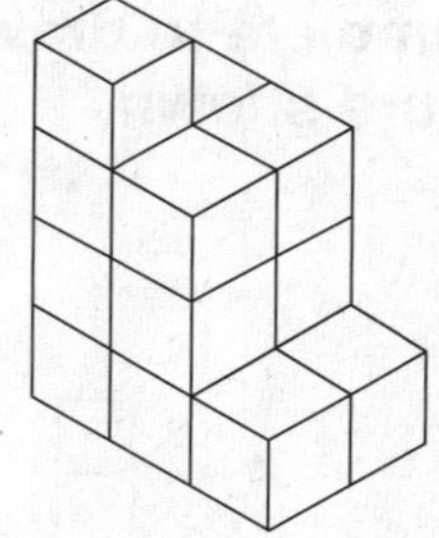

Name ______________________ Date ____________ Class ____________

LESSON **8-4**

Problem Solving

Drawing Three-Dimensional Figures

Write the correct answer.

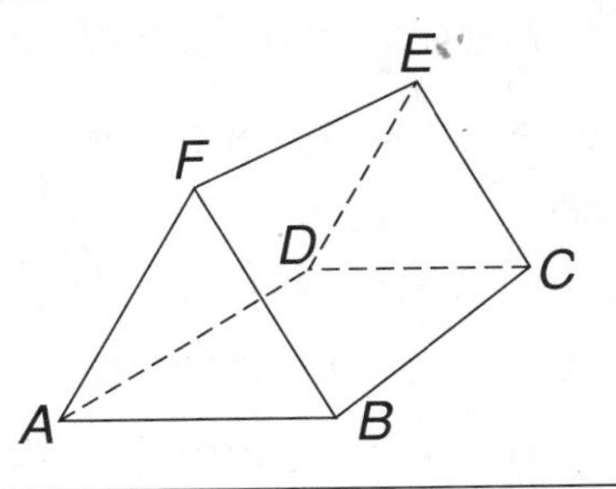

1. Mitch used a triangular prism in his science experiment. Name the vertices, edges, and faces of the triangular prism shown at the right.

____

____

____

2. Amber used cubes to make the model shown below of a sculpture she wants to make. Draw the front, side, and top views of the model.

Choose the letter of the best answer.

3. Which is the front view of the figure shown at the left? ____

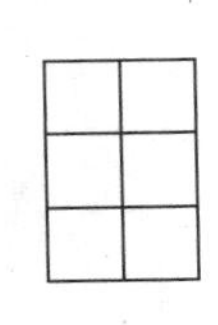

A

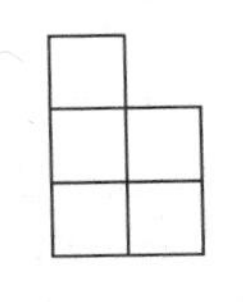

B

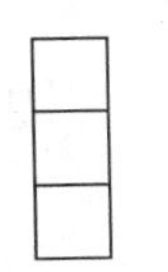

C

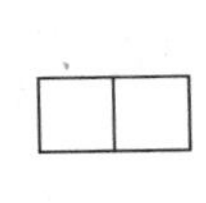

D

4. Which is **not** an edge of the figure shown at the right?

F $\overline{AB}$ **H** $\overline{EF}$

G $\overline{EH}$ **J** $\overline{BD}$

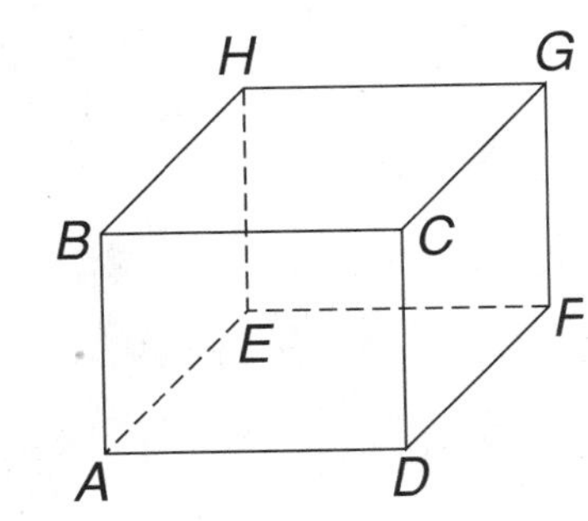

Name ______________________ Date __________ Class __________

LESSON 8-4

Reading Strategies

Using a Model

This cube is made up of faces, edges, and vertices.

- A **face** is a flat surface.
- An **edge** is where two faces meet.
- A **vertex** is where three or more edges meet.

vertex

edge

face

When you draw a cube on a flat surface, it is impossible to see all six faces.

A **net** is like a wrapper for the cube. When it is unwrapped and lying flat, you can see all the faces.

Every solid figure has a net.

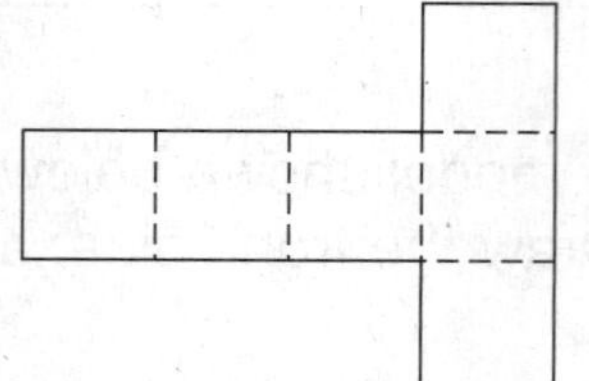

For Exercises 1–4, match each solid figure with its net. Write a, b, c, or d.

a.

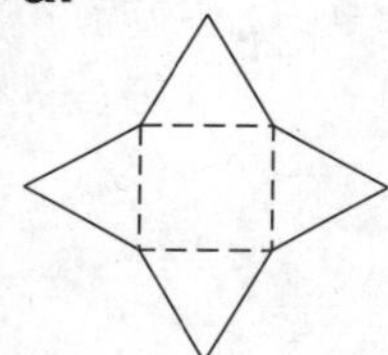

b.

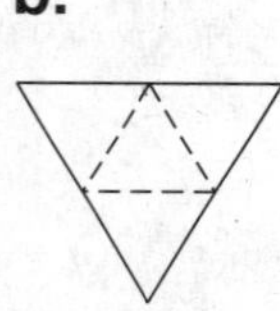

c.

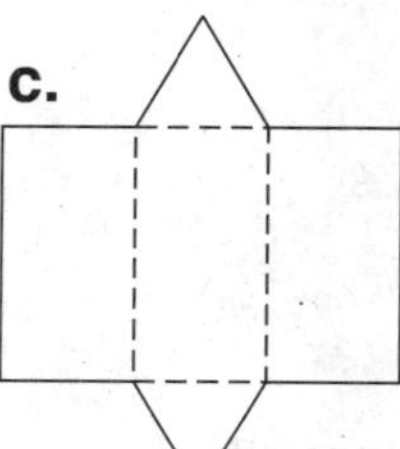

d.

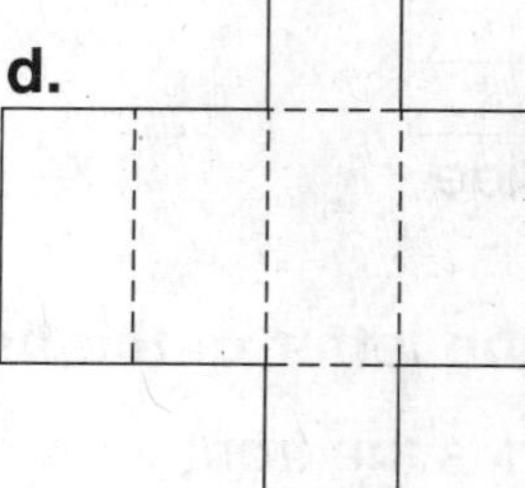

1.

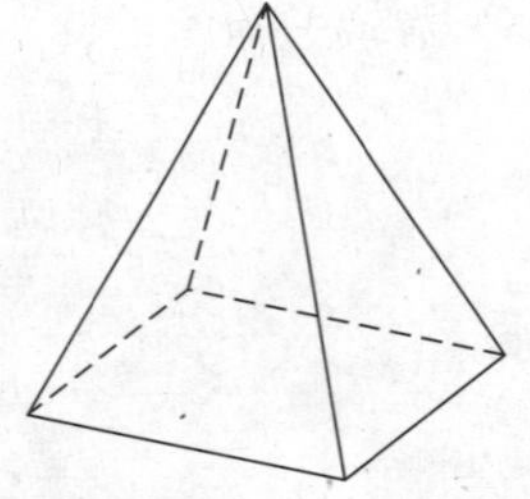

____

2.

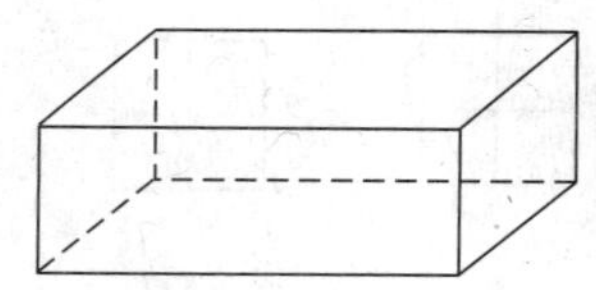

____

3.

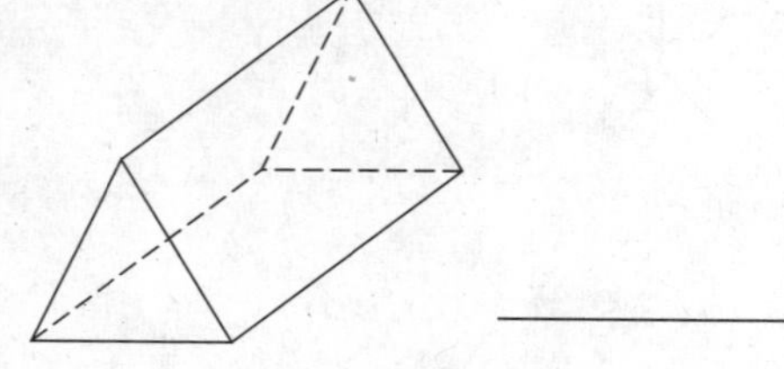

____

4.

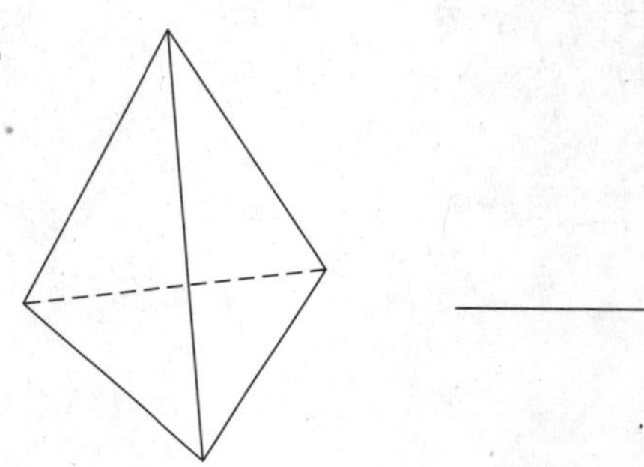

____

Name ______________________ Date ____________ Class ____________

LESSON **8-4**

Puzzles, Twisters & Teasers

A Three-Dimensional State!

Circle words from the list in the word search (horizontally, vertically or diagonally). Then find a word that answers the riddle. Circle it and write it on the line.

face	edge	vertex	vertices	top	front
side	dimension	view	cube	perspective	orthogonal

What goes up but never comes down? ______________________

Name ______________________ Date __________ Class __________

LESSON 8-5

Practice A

Volume of Prisms and Cylinders

Find the volume to the nearest tenth of a unit. Prism: $V = Bh$. Cylinder: $V = \pi r^2 h$. Use 3.14 for π.

1.

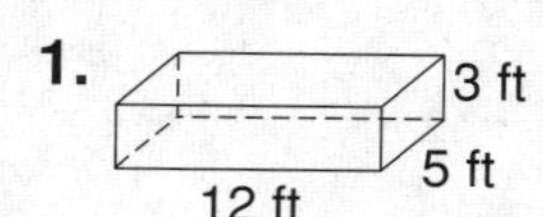

____

2.

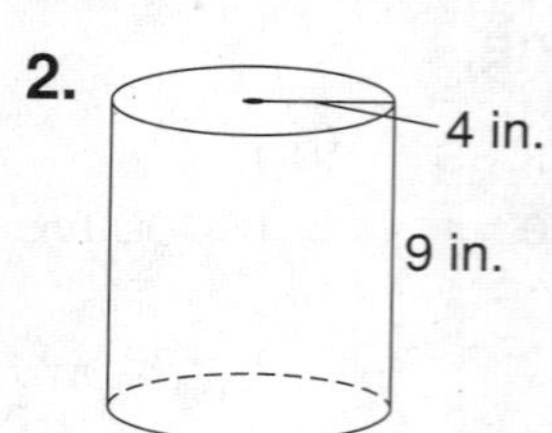

____

3.

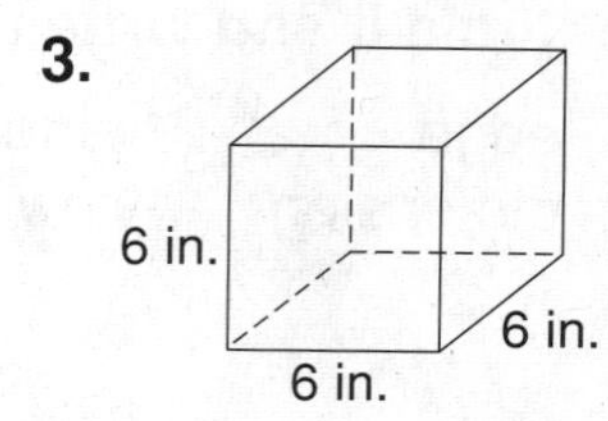

____

4.

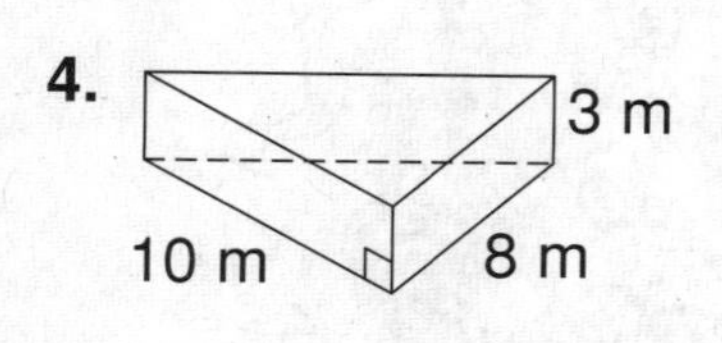

____

5.

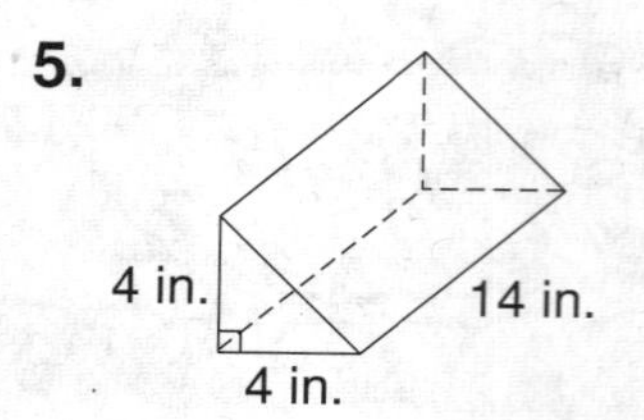

____

6.

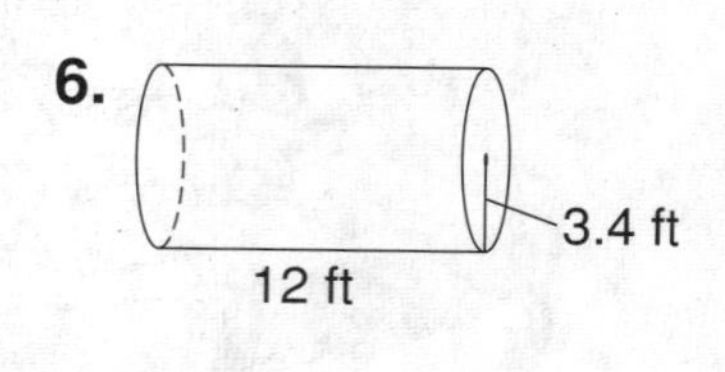

____

7.

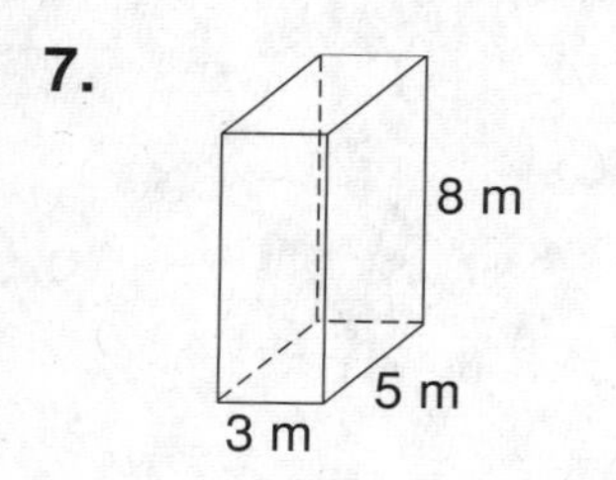

____

8.

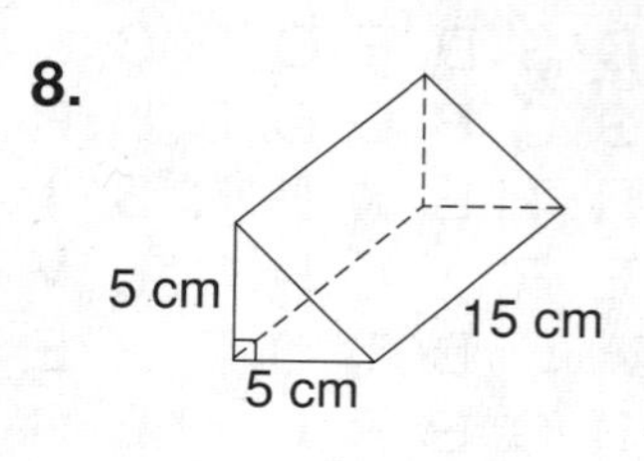

____

9.

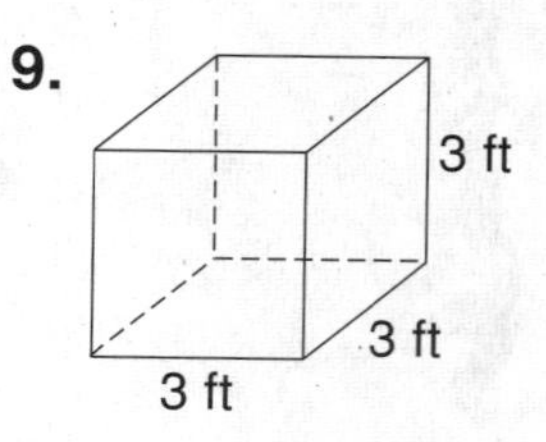

____

10. A rectangular box measures 6 ft by 8 ft by 2 ft. Explain whether doubling a side from 6 ft to 12 ft would double the volume of the box.

____

____

11. A can of vegetables is 4.5 in. high and has a diameter of 3 in. Find the volume of the can to the nearest tenth of a unit. Use 3.14 for π.

____

12. A telephone pole is 30 ft tall with a diameter of 12 in. Jacob is making a replica of a telephone pole and wants to fill it with sand to help it stand freely. Find the volume of his model, which has a height of 30 in. and a diameter of 1 in., to the nearest tenth of a unit. Use 3.14 for π.

____

Name ______________________ Date ____________ Class ____________

LESSON 8-5

Practice B

Volume of Prisms and Cylinders

Find the volume of each figure to the nearest tenth. Use 3.14 for π.

1.

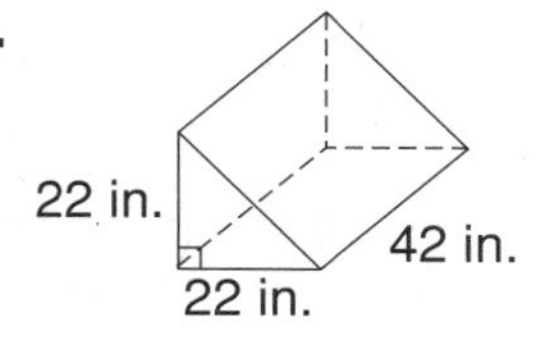

____

2.

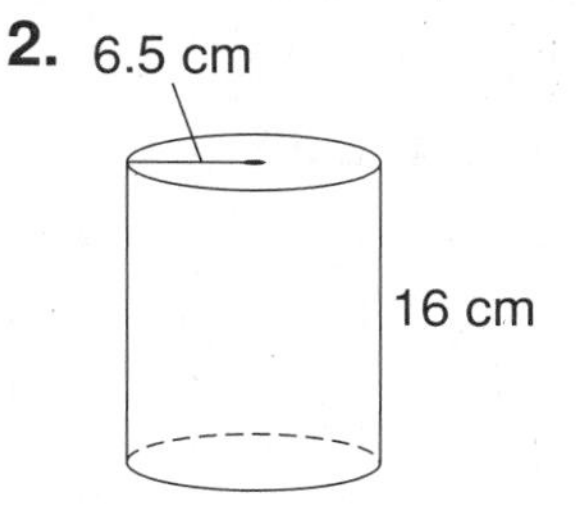

____

3.

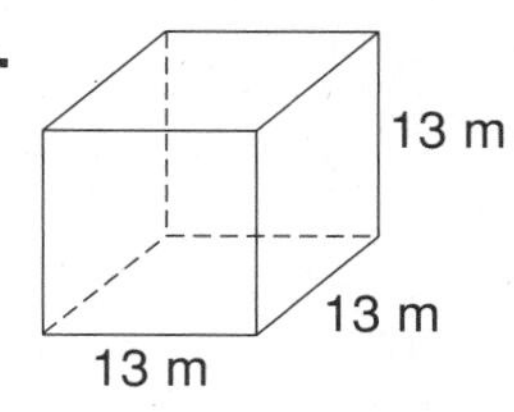

____

4.

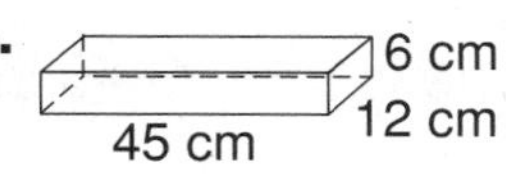

____

5.

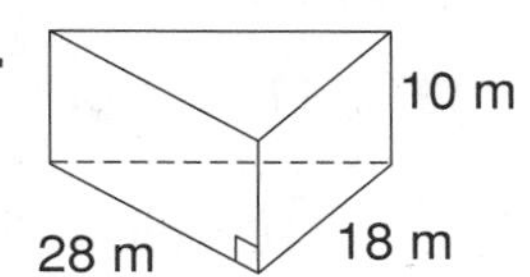

____

6.

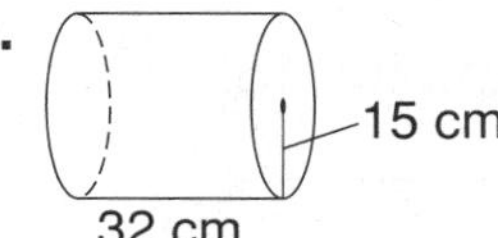

____

7.

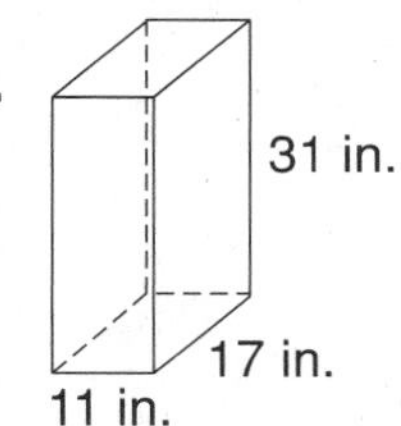

____

8.

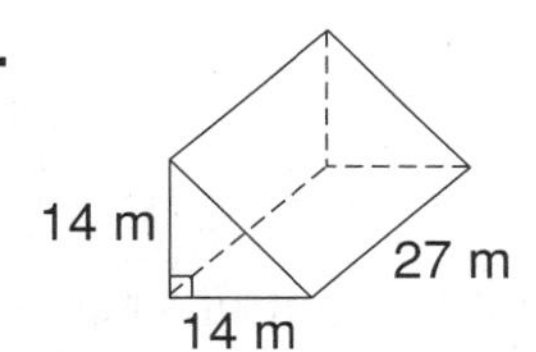

____

9.

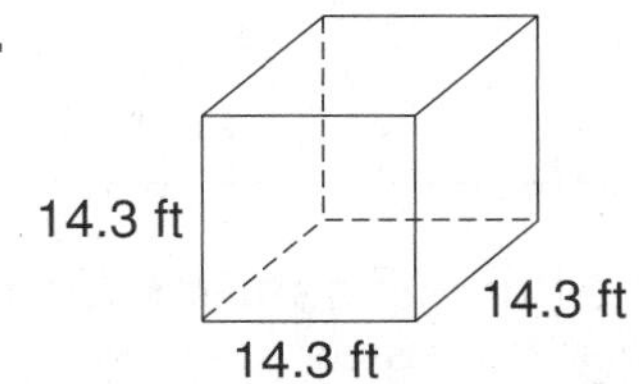

____

10. A cylinder has a radius of 6 ft and a height of 25 ft. Explain whether tripling the height will triple the volume of the cylinder.

____

____

11. Contemporary American building bricks are rectangular blocks with the standard dimensions of about 5.7 cm by 9.5 cm by 20.3 cm. What is the volume of a brick to the nearest tenth of a unit?

____

12. Ian is making candles. His cylindrical mold is 8 in. tall and has a base with a diameter of 3 in. Find the volume of a finished candle to the nearest tenth of a unit.

____

Name ______________________ Date __________ Class __________

LESSON 8-5

Practice C

Volume of Prisms and Cylinders

Find the volume of each figure to the nearest tenth. Use 3.14 for π.

1.

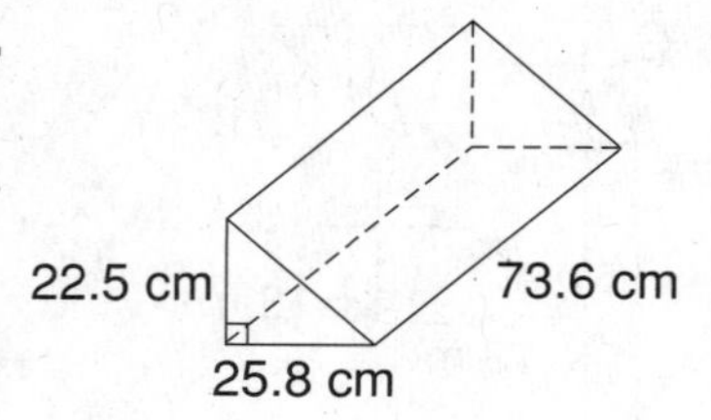

2.

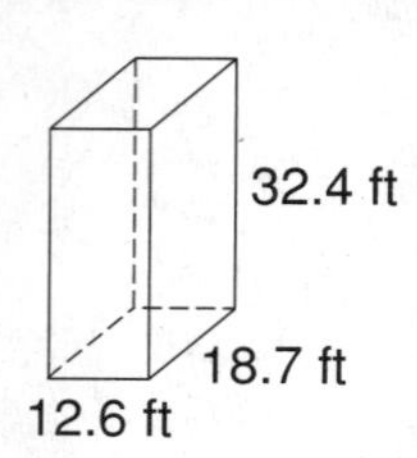

3.

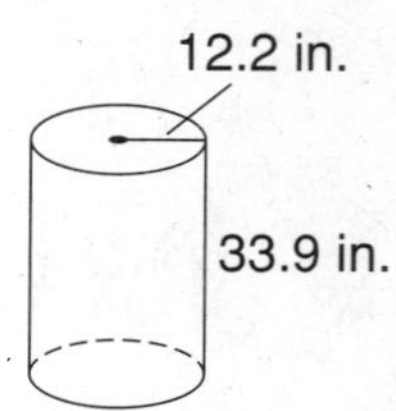

______________ ______________ ______________

Find the missing measure to the nearest tenth. Use 3.14 for π.

4. volume is 1152 ft^3

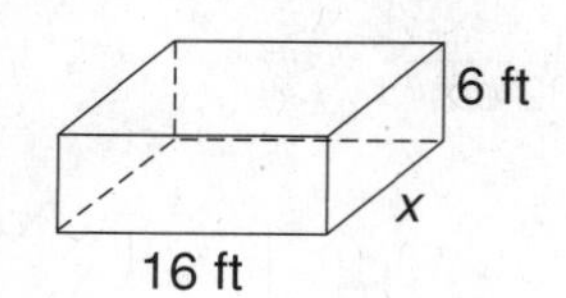

5. volume is 1132.8 cm^3

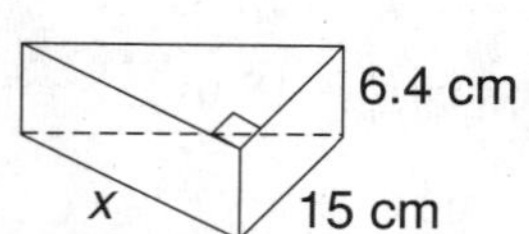

6. volume is 39,193.48 in^3

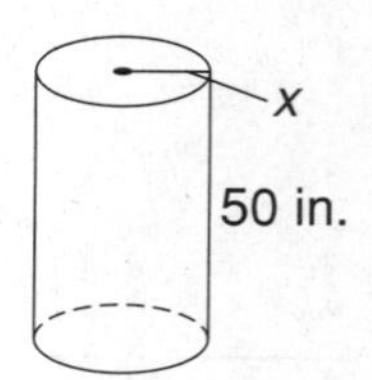

______________ ______________ ______________

7. Atmo's Theater's large-sized popcorn container is a cylinder with a diameter of 9 in. and is 6 in. tall. The medium-sized container has a diameter of 6 in. and is 8 in. tall. Find the difference in the volumes. ______________

8. The Blackmans have a new 10-ft by 20-ft rectangular swimming pool in their backyard. It has a uniform depth of 5 ft. What is the volume of water needed to fill the pool? ______________

9. Donna decorated a three-tiered cake for her parent's anniversary party. The bottom tier had a 12-in. diameter, the middle tier, a 9-in. diameter, and the top tier, a 6-in. diameter. Each tier was 4 in. tall. What was the total volume of the cake to the nearest tenth of a unit? ______________

10. A company produces 100,000 boxes of cereal with dimensions of 12 in. (height) by 8 in. by 1.5 in. The manufacturer is trying to cut production cost per box so the height of the box is decreased by one inch. Approximately how many more boxes of cereal can be produced in a month with the new dimensions using the same amount of cereal? Round the answer to the nearest integer. ______________

Name ______________________ Date __________ Class __________

LESSON **8-5**

Reteach

Volume of Prisms and Cylinders

Volume = number of cubic units inside a solid figure

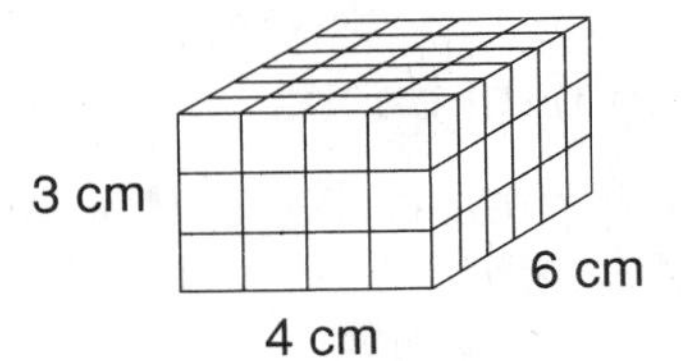

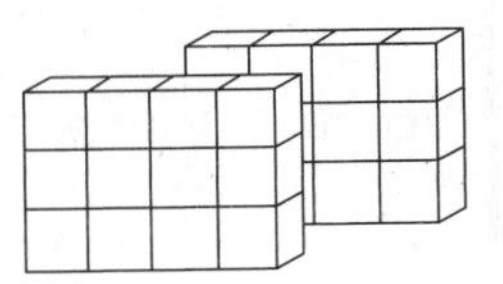

To find the volume of this solid figure:

Count the number of cubic centimeters in one "slice" of the figure.

$4 \times 3 = 12$

Multiply by the number of "slices." $12 \times 6 = 72$ cm^3

Complete to find the volume of each solid figure.

1.

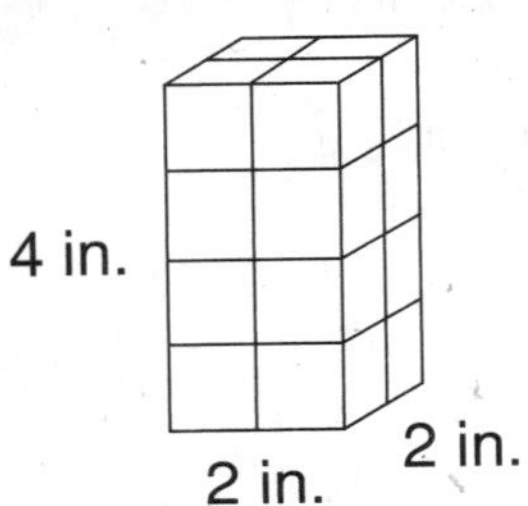

number in^3 in a slice

= ______ × ______ = ______

number of slices = ______

volume = ______ in^3

2.

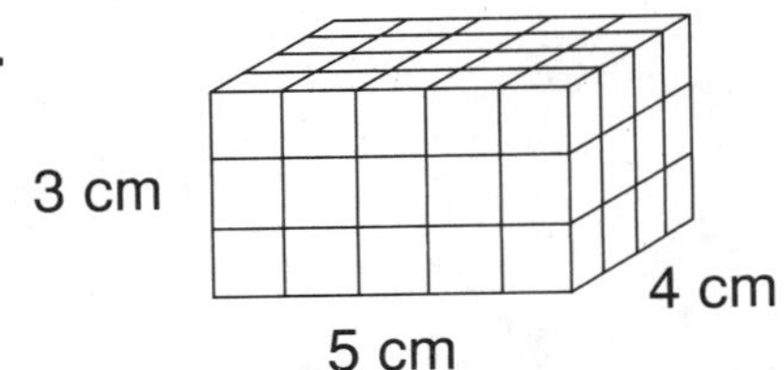

number cm^3 in a slice

= __________ = ______

number of slices = ______

volume = ______ cm^3

3.

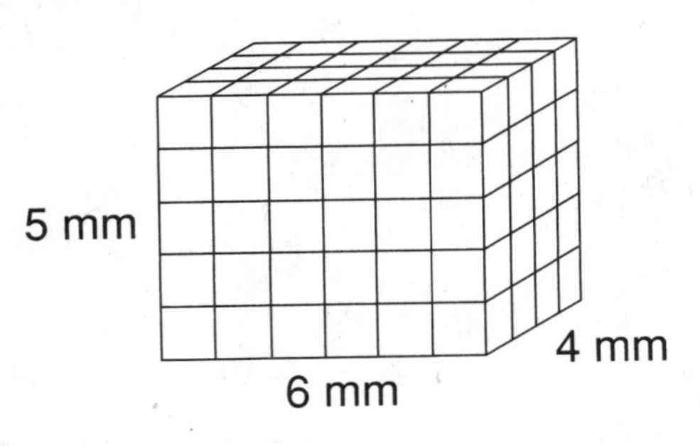

number mm^3 in a slice

= __________ = ______

number of slices = ______

volume = ______ mm^3

Name ______________________ Date ____________ Class ____________

LESSON 8-5

Reteach

Volume of Prisms and Cylinders (continued)

Prism: solid figure named for the shape of its two congruent bases

Volume V of a prism = area of base B × height h

$\boldsymbol{V = Bh}$

$V = (6 \times 4)5$

$V = 24(5)$

$V = 120 \text{ in}^3$

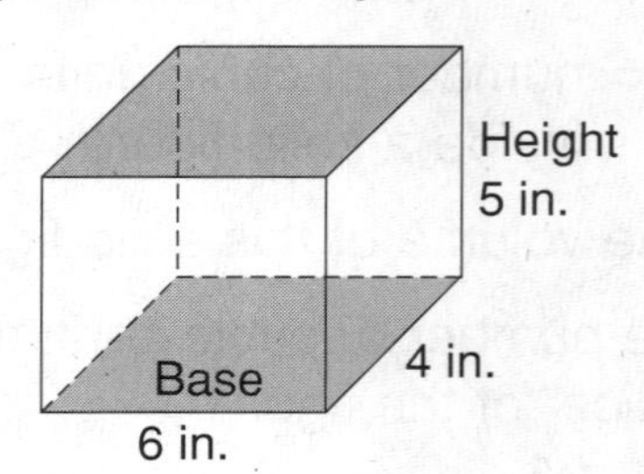

Rectangular Prism

Complete to find the volume of each prism.

4. rectangular prism

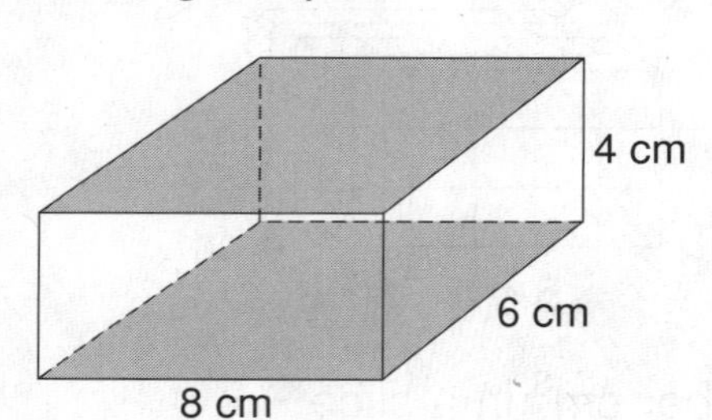

base is a rectangle

$V = Bh$

$V =$ (____ × ____) × ____

= ____ cm^3

5. cube

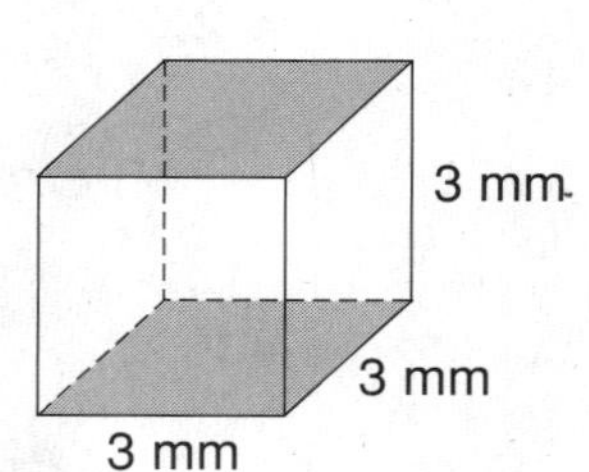

base is a ____________

$V = Bh$

$V =$ (____ × ____) × ____

= ____ mm^3

6. triangular prism

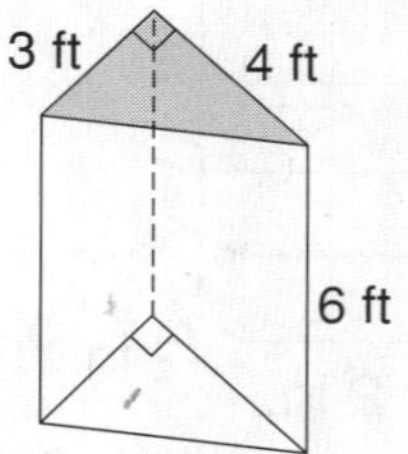

base is a ____________

$V = \frac{1}{2}Bh$

$V = \frac{1}{2}$(____ × ____)

× ____ = ____ ft^3

Cylinder: solid figure with a circular base

Volume V of a cylinder = area of base B × height h

$\boldsymbol{V = Bh}$

$V = (\pi \times 4^2)7 = (16\pi)7$

$V = 112\pi$

$V \approx 112(3.14) \approx 351.7 \text{ units}^3$

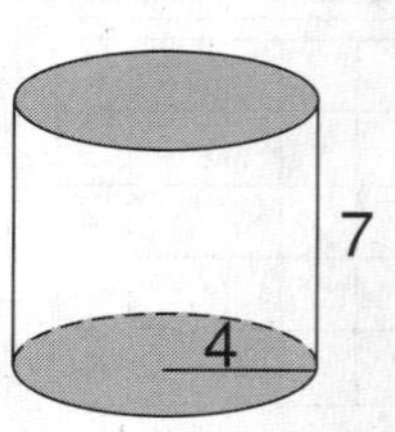

Complete to find the volume of the cylinder.

7.

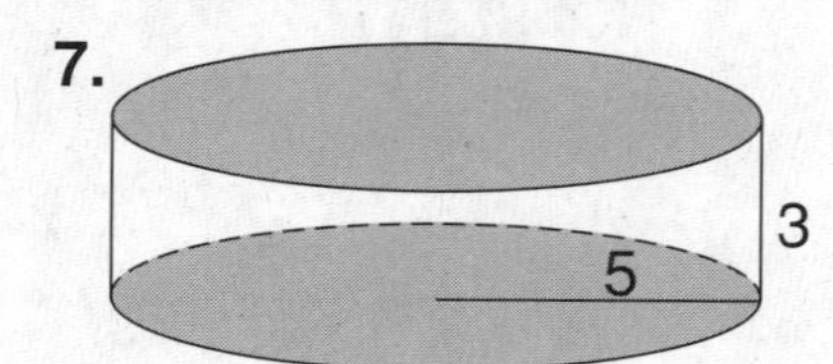

$V = Bh$

$V = (\pi \times$ ______$) \times$ ______

$V =$ ______ $\approx$ ______ units^3

Name ______________________ Date ____________ Class ____________

LESSON 8-5

Challenge

Looking Askance

So far, you have considered prisms in which the outside edges are perpendicular to the plane of the base.

Now, you will consider prisms in which the outside edges are not perpendicular to the plane of the base.

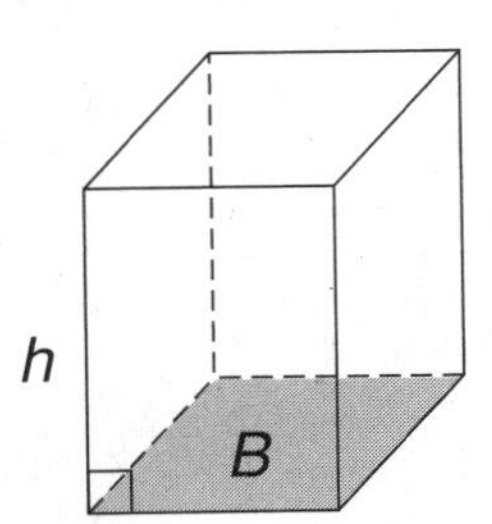

Right Prism

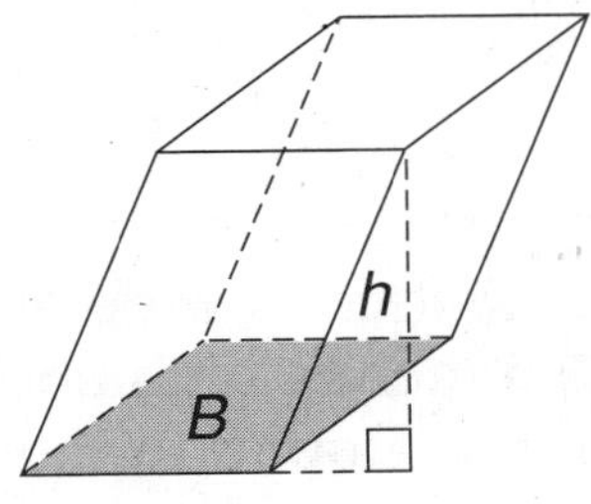

Oblique Prism

1. Explain why these two prisms have the same volume.

____

A prism can have any polygon as its base. Consider an oblique prism with a base that is a parallelogram.

To find the volume of this prism, first look at parallelogram *JKLM* which is the base of the prism.

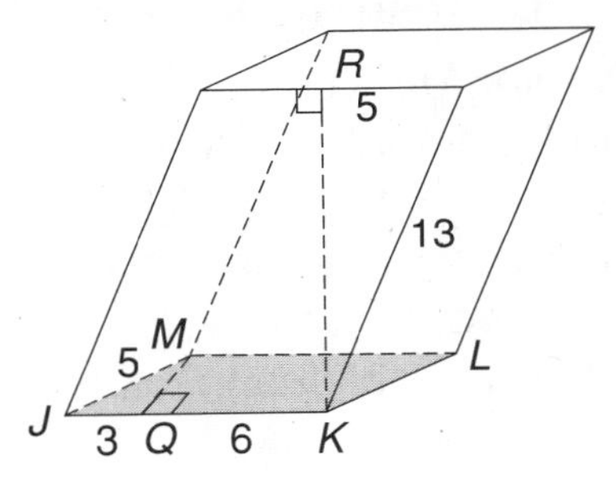

2. How long is *JK*, the base of parallelogram *JKLM?*

$JK =$ ______________________

3. Find *MQ*, the height of parallelogram *JKLM*. Explain your method.

$MQ =$ ______________________

4. What is the area of the base of the prism?

$B =$ ______________________

5. Find *KR*, the height of the prism. Explain your method.

$KR =$ ______________________

6. Find the volume of the prism.

$V =$ ______________________

Name ______________________ Date __________ Class __________

LESSON 8-5

Problem Solving

Volume of Prisms and Cylinders

Round to the nearest tenth. Write the correct answer.

1. A contractor pours a sidewalk that is 4 inches deep, 1 yard wide, and 20 yards long. How many cubic yards of concrete will be needed? (Hint: 36 inches = 1 yard.)

____

2. A refrigerator has inside measurements of 50 cm by 118 cm by 44 cm. What is the capacity of the refrigerator?

____

A rectangular box is 2 inches high, 3.5 inches wide and 4 inches long. A cylindrical box is 3.5 inches high and has a diameter of 3.2 inches. Use 3.14 for π. Round to the nearest tenth.

3. Which box has a larger volume?

____

4. How much bigger is the larger box?

____

Use 3.14 for π. Choose the letter for the best answer.

5. A child's wading pool has a diameter of 5 feet and a height of 1 foot. How much water would it take to fill the pool? Round to the nearest gallon. (Hint: 1 cubic foot of water is approximately 7.5 gallons.)

   A 79 gallons

   B 589 gallons

   C 59 gallons

   D 147 gallons

6. How many cubic feet of air are in a room that is 15 feet long, 10 feet wide and 8 feet high?

   F 33 ft^3

   G 1200 ft^3

   H 1500 ft^3

   J 3768 ft^3

7. How many gallons of water will the water trough hold? Round to the nearest gallon. (Hint: 1 cubic foot of water is approximately 7.5 gallons.)

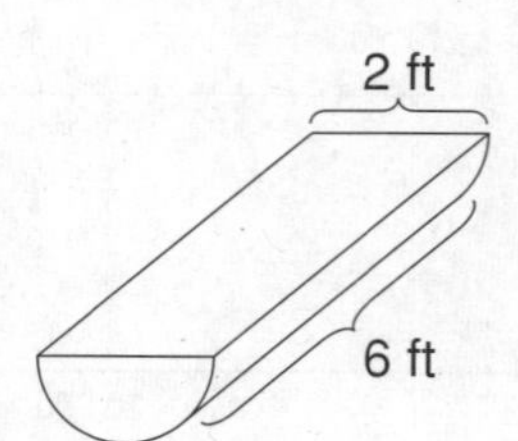

   A 19 gallons **C** 141 gallons

   B 71 gallons **D** 565 gallons

8. A can has diameter of 9.8 cm and is 13.2 cm tall. What is the capacity of the can? Round to the nearest tenth.

   F 203.1 cm^3

   G 995.2 cm^3

   H 3980.7 cm^3

   J 959.2 cm^3

Name ______________________ Date __________ Class __________

LESSON **8-5**

Reading Strategies

Using a Model

Volume is the amount of space a solid figure occupies.
Volume is measured in cubic units.

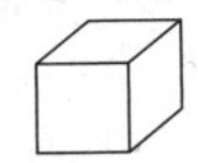

One cubic unit

A **rectangular prism** is a three-dimensional figure. The length, width, and height are the measures that make up the three dimensions.

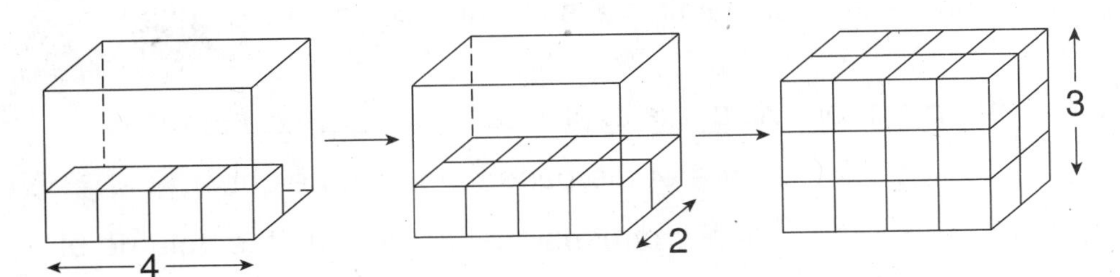

You can find the volume of a rectangular prism by multiplying the length times the width times the height.

The formula for the volume of a rectangular prism is:
$V = \ell \cdot w \cdot h$.

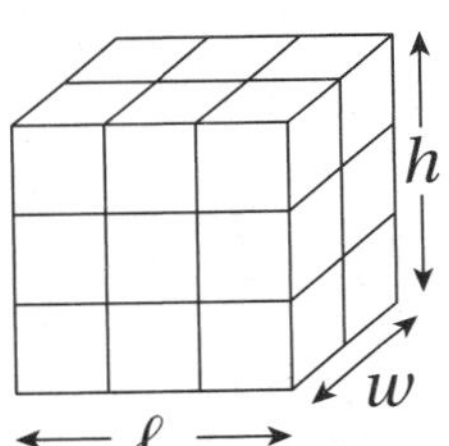

Use the rectangular prism to complete Exercises 1–6.

1. How long is the prism? ____________

2. How wide is the prism? ____________

3. What is the height of the prism? ____________

4. Write the formula you will use to find the volume of this prism?

____

5. Rewrite the formula with values for each dimension of the prism.

____

6. What is the volume of this prism?

____

Name ______________________ Date __________ Class __________

LESSON 8-5

Puzzles, Twisters & Teasers

Turn Up the Volume!

Across

2. A __________ is a three-dimensional figure named for the shape of its bases.

4. The volume of a prism or cylinder is expressed in __________ units.

6. The volume of a __________ prism can be written as $V = \ell wh$.

9. The volume of a prism is the area of the __________ times the height.

Down

1. The __________ of a three-dimensional figure is the number of cubic units needed to fill it.

3. The area of the base of a prism or cylinder is expressed in __________ units.

5. The volume of a __________ is the area of the base times the height.

7. To find the volume of a __________ three-dimensional figure, find the volume of each part and add the volumes together.

8. If all six faces of a rectangular prism are squares, it is a __________ .

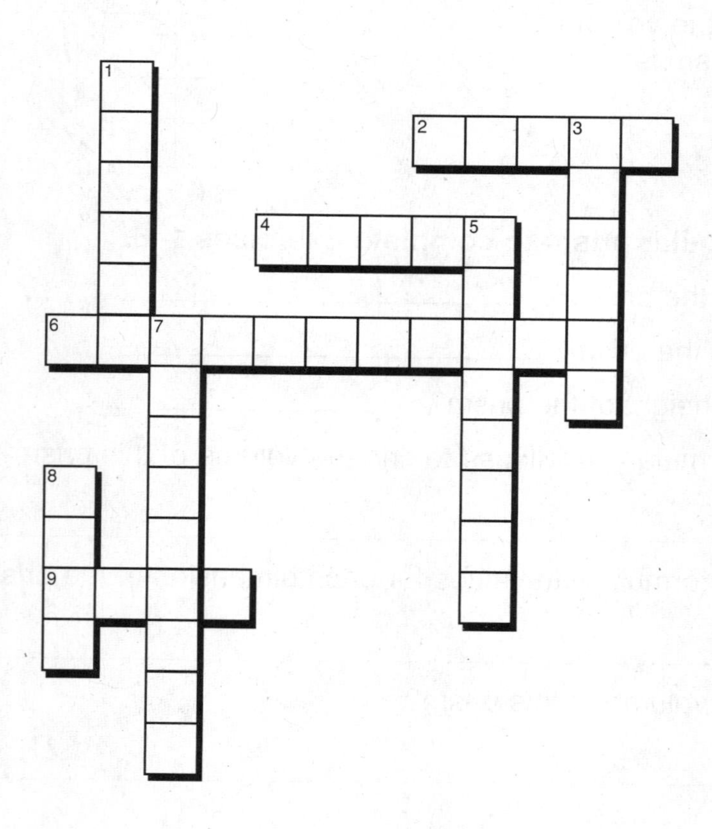

Name ______________________ Date ___________ Class ___________

LESSON 8-6

Practice A

Volume of Pyramids and Cones

Find the volume of each figure to the nearest tenth. Use 3.14 for π.

Pyramid: $V = \frac{1}{3}Bh$. Cone: $V = \frac{1}{3}\pi r^2 h$. Use 3.14 for π.

1.

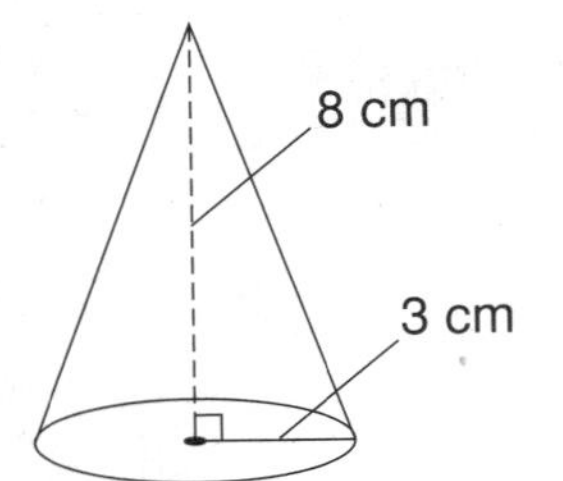

____

2.

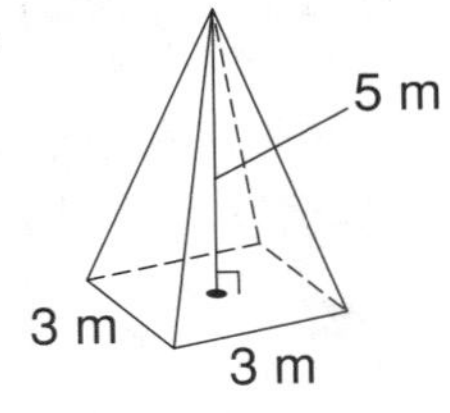

____

3.

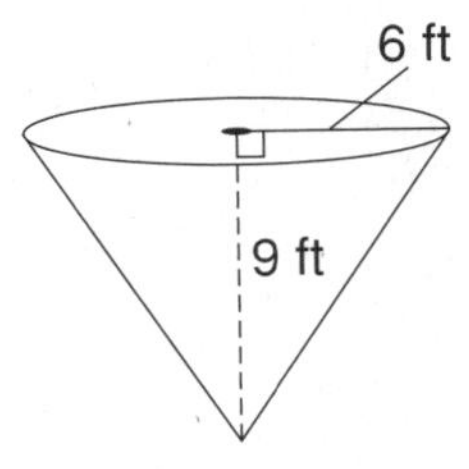

____

4.

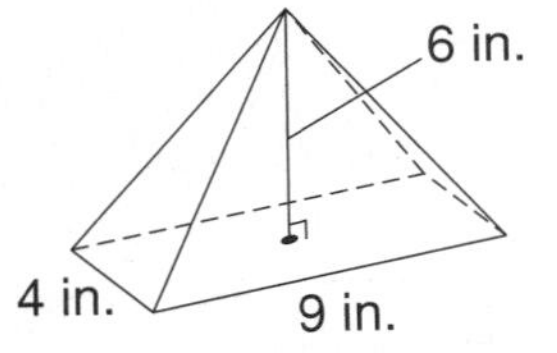

____

5.

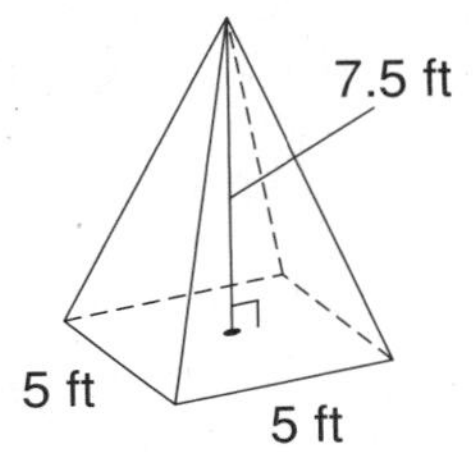

____

6.

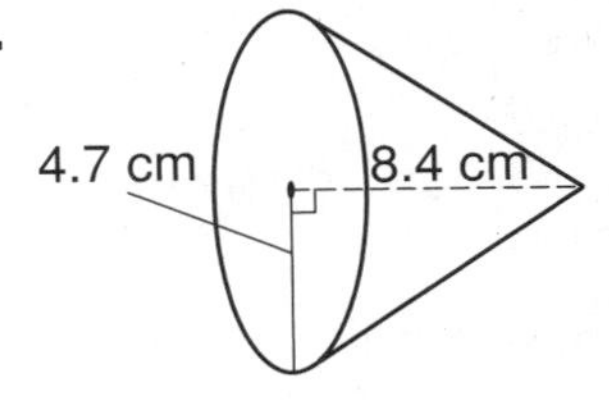

____

7. The base of a regular pyramid has an area of 12 ft^2. The height of the pyramid is 8 ft. Find the volume. ________________

8. The radius of a cone is 4.9 cm and its height is 10 cm. Find the volume of the cone to the nearest tenth. Use 3.14 for π. ________________

9. Find the volume of a rectangular pyramid if the height is 6 m and the base sides are 7 m and 5 m. ________________

10. The mold for an ice cone has a diameter of 4 in. and is 5 in. deep. Use a calculator to find the volume of the ice cone mold to the nearest hundredth. ________________

11. A square pyramid has a height 6 cm and a base that measures 5 cm on each side. Explain whether doubling the height would double the volume of the pyramid.

____

____

____

Name ______________________ Date __________ Class __________

LESSON 8-6

Practice B

Volume of Pyramids and Cones

Find the volume of each figure to the nearest tenth. Use 3.14 for π.

1.

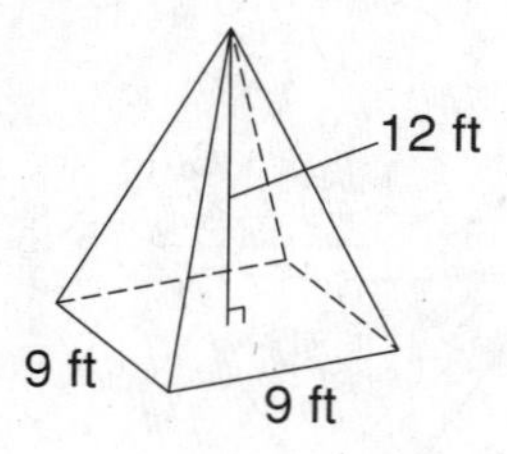

____

2.

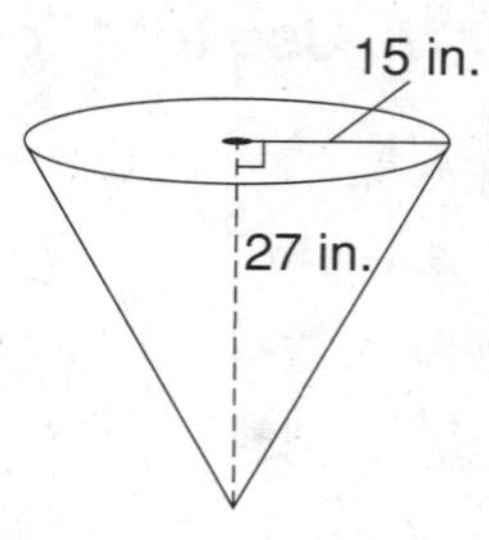

____

3.

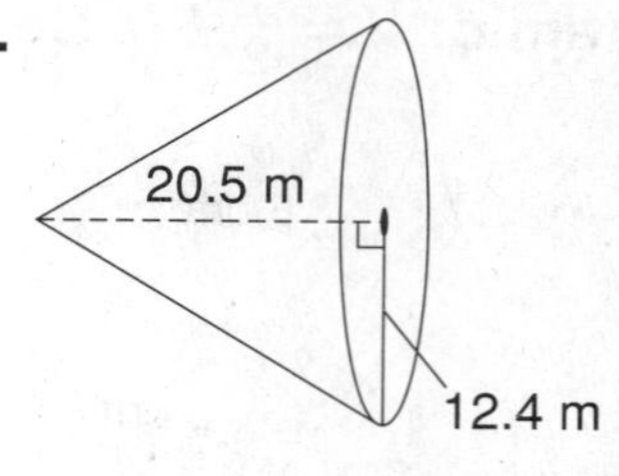

____

4.

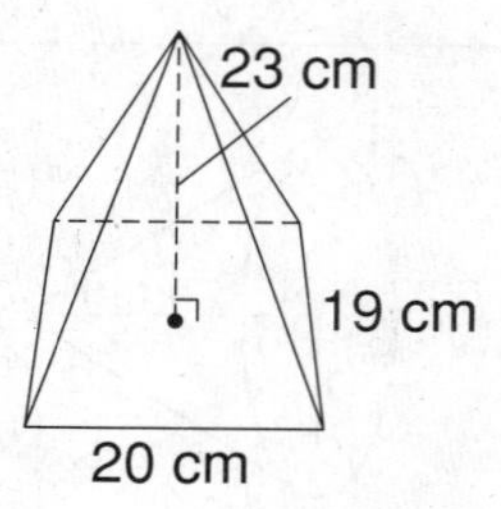

____

5.

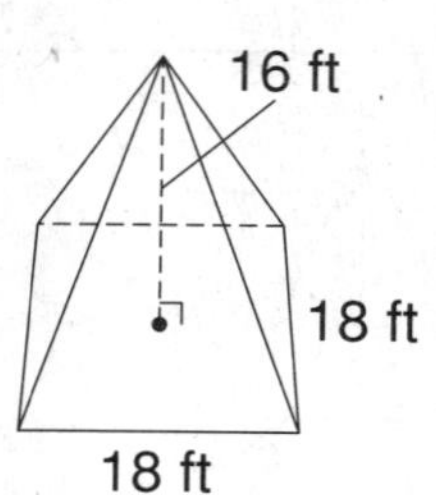

____

6.

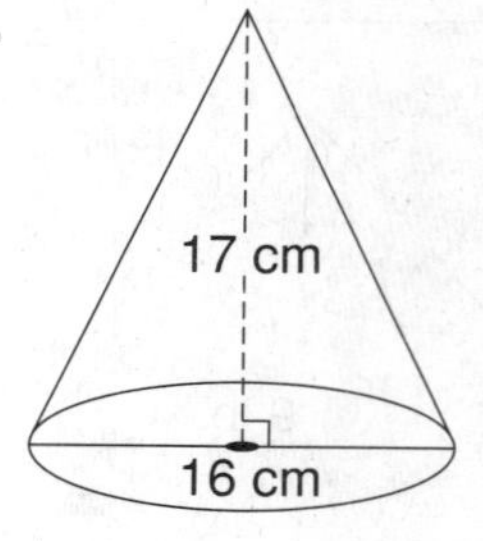

____

7. The base of a regular pyramid has an area of 28 in^2. The height of the pyramid is 15 in. Find the volume. ______________

8. The radius of a cone is 19.4 cm and its height is 24 cm. Find the volume of the cone to the nearest tenth. ______________

9. Find the volume of a rectangular pyramid if the height is 13 m and the base sides are 12 m and 15 m. ______________

10. A funnel has a diameter of 9 in. and is 16 in. deep. Use a calculator to find the volume of the funnel to the nearest hundredth. ______________

11. A square pyramid has a height 18 cm and a base that measures 12 cm on each side. Explain whether tripling the height would triple the volume of the pyramid.

____

____

____

Name ______________________ Date __________ Class __________

LESSON 8-6

Practice C

Volume of Pyramids and Cones

Find the volume of each figure to the nearest tenth. Use 3.14 for π.

1.

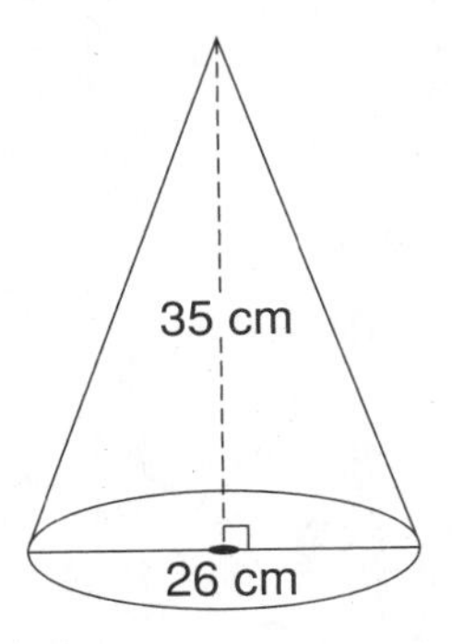

____

2.

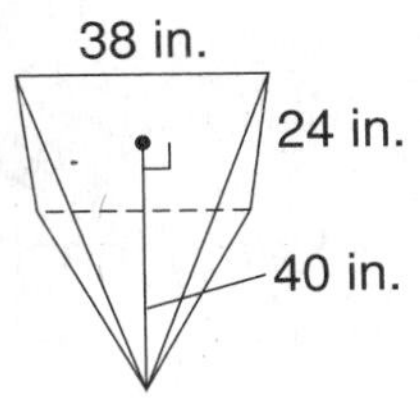

____

3.

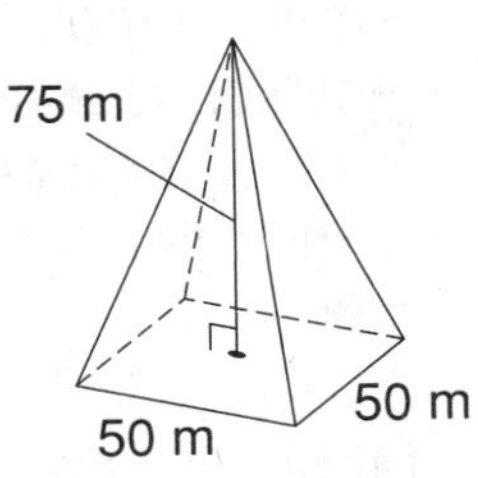

____

Find the missing measure to the nearest tenth. Use 3.14 for π.

4. rectangular pyramid:
   base length = 15 m
   base width = ?
   height = 21 m
   volume = 2415 m^3

   ____

5. triangular pyramid:
   base width = 8 cm
   base height = 18 cm
   height = ?
   volume = 624 cm^3

   ____

6. A cone has diameter of 24 ft and height of 15 ft. How many times will the volume of the cone fill a cylinder with radius of 18 ft and a height of 25 ft? Round your answer to the nearest whole number. ______________

7. Find the volume of the figure to the nearest tenth.

   ____

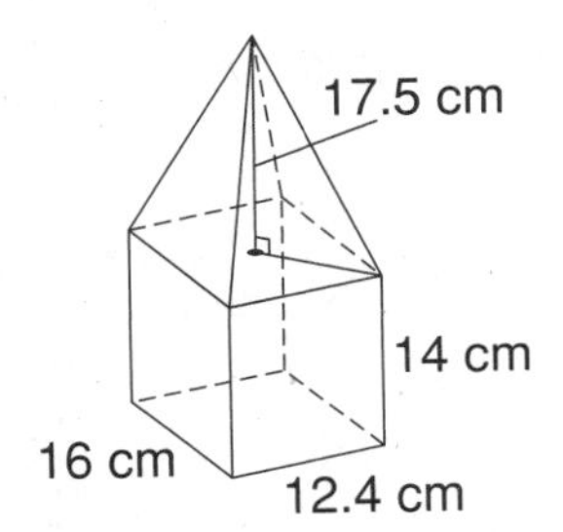

8. Find the volume of the figure to the nearest tenth.

   ____

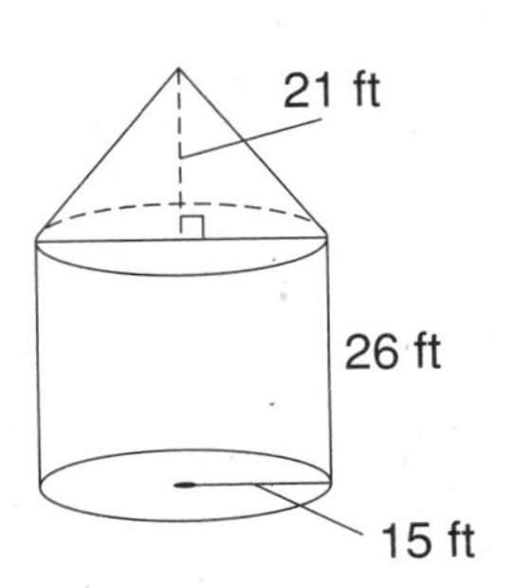

Name ______________________ Date ____________ Class ____________

LESSON 8-6

Reteach

Volume of Pyramids and Cones

Pyramid: solid figure named for the shape of its base, which is a polygon; all other faces are triangles

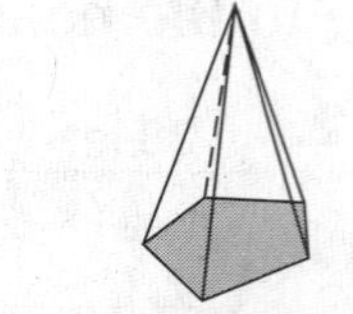

Pentagonal Pyramid

This rectangular pyramid and rectangular prism have congruent bases and congruent heights.

Volume of Pyramid = $\frac{1}{3}$ Volume of Prism

$$V = \frac{1}{3}Bh$$

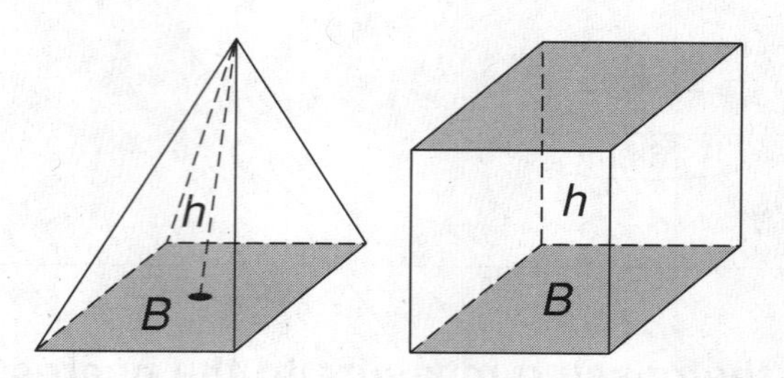

Complete to find the volume of each pyramid.

1. square pyramid

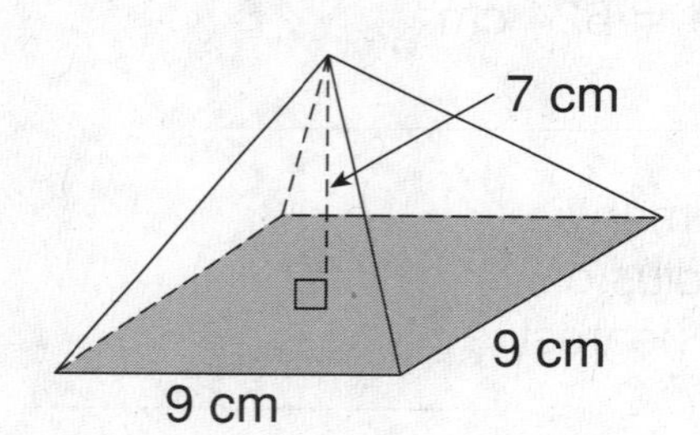

base is a ____________

$V = \frac{1}{3}Bh$

$V = \frac{1}{3}$ (area of square) × h

$V = \frac{1}{3}$ (____ × ____) × ____

$V = \frac{1}{3}$ (____) × ____

$V =$ ________ cm^3

2. rectangular pyramid

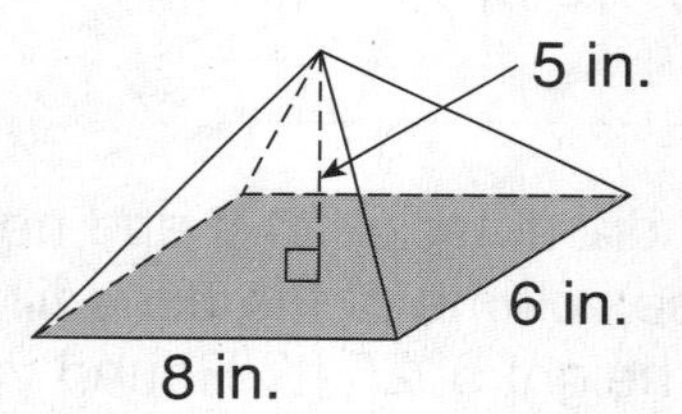

base is a ____________

$V = \frac{1}{3}Bh$

$V = \frac{1}{3}$ (area of rectangle) × h

$V = \frac{1}{3}$ (____ × ____) × ____

$V = \frac{1}{3}$ (____) × ____

$V =$ ________ in^3

Name ______________________ Date ___________ Class ___________

LESSON 8-6

Reteach

Volume of Pyramids and Cones (continued)

Cone: solid figure with a circular base

This cone and cylinder have congruent bases and congruent heights.

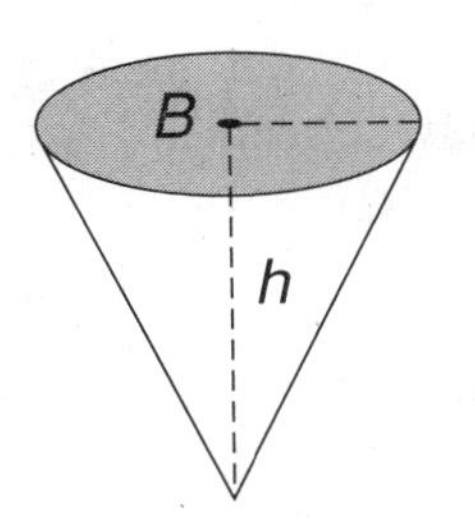

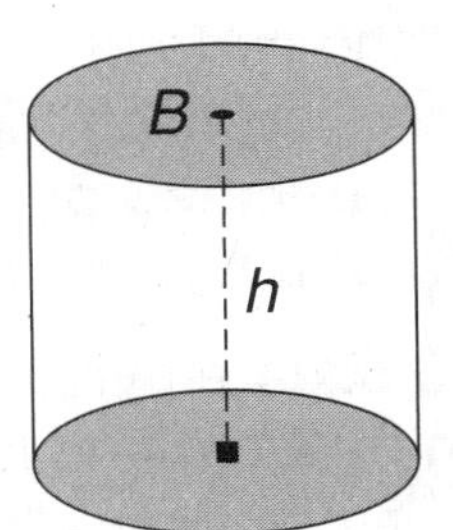

Volume of Cone = $\frac{1}{3}$ Volume of Cylinder

$$V = \frac{1}{3}Bh$$

Complete to find the volume of each cone.

3.

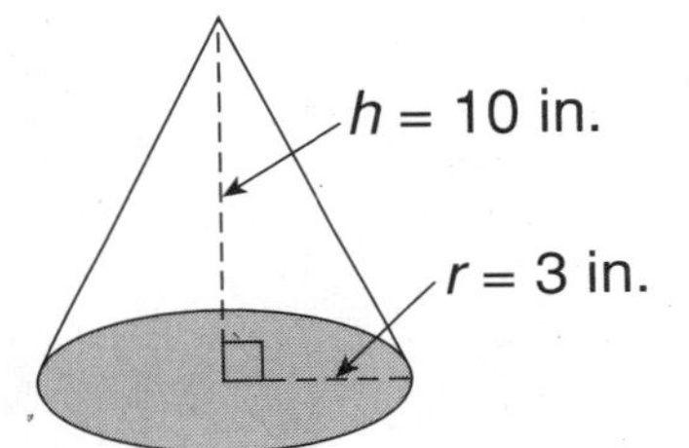

radius r of base = ______ in.

$V = \frac{1}{3}Bh$

$V = \frac{1}{3}(\pi r^2)h$

$V = \frac{1}{3}(\pi \times$ ______$) \times$ ______

$V = \frac{1}{3}($______$) \times$ ______

$V =$ ______ $\times$ ______

$V =$ ______

$V \approx$ ______ $\times 3.14$

$V \approx$ ____________ in^3

4.

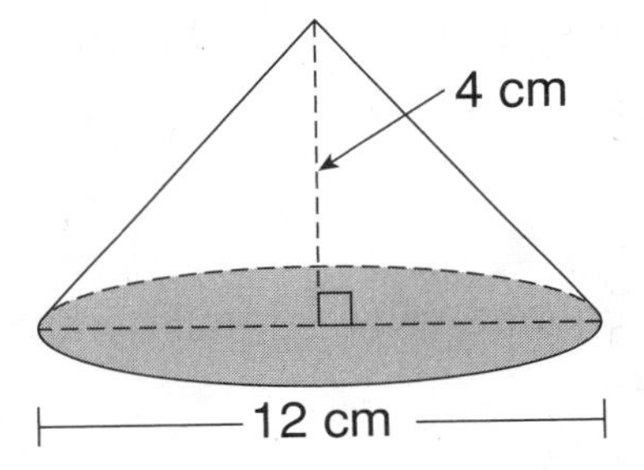

radius $r = \frac{1}{2}$ diameter = ______ cm

$V = \frac{1}{3}Bh$

$V = \frac{1}{3}(\pi r^2)h$

$V = \frac{1}{3}(\pi \times$ ______$) \times$ ______

$V = \frac{1}{3}($______$) \times$ ______

$V =$ ______ $\times$ ______

$V =$ ______

$V \approx$ ______ $\times 3.14$

$V \approx$ ____________ cm^3

Name ______________________________ Date ____________ Class ____________

LESSON 8-6

Challenge

Take a Little Off the Top

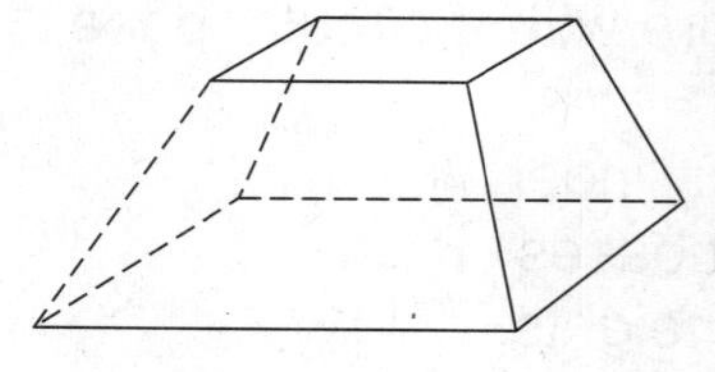

When a pyramid is cut by a plane parallel to its base, the part of the surface between and including the two planes is called a **truncated pyramid** or **frustum of a pyramid**.

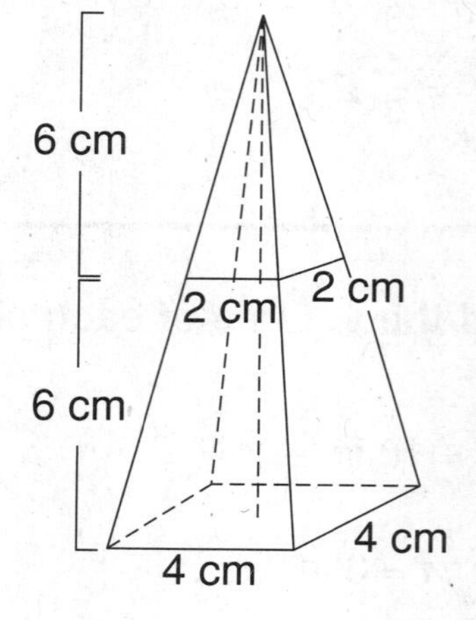

Consider a square pyramid of base 4 cm by 4 cm and height 12 cm. As shown, a plane parallel to the base cuts the pyramid halfway down.

1. Use the familiar formula $V = \frac{1}{3}Bh$ where B = area of base and h = height to find the volume of the large pyramid.

$V_{\text{large pyramid}}$ = ____________________ cm^3

2. Find the volume of the small pyramid at the top.

$V_{\text{small pyramid}}$ = ____________________ cm^3

3. Use your results to find the volume.

$V_{\text{truncated pyramid}}$ = ____________________ cm^3

A papyrus roll written in ancient Egypt around 1890 B.C.E.—now known as the Moscow Papyrus because it was brought to Russia in 1893—gives a formula for finding the volume of a truncated pyramid.

$$V_{\text{truncated pyramid}} = \frac{1}{3}h(a^2 + ab + b^2)$$

where
h = height
a = edge length of top base
b = edge length of bottom base

4. Use the ancient formula to find the volume of the truncated pyramid. Compare your results from exercises 3 and 4.

$V_{\text{truncated pyramid}}$ = ____________________ cm^3

Name ______________________ Date ____________ Class ____________

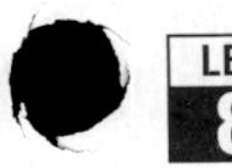

LESSON **8-6**

Problem Solving

Volume of Pyramids and Cones

Round to the nearest tenth. Use 3.14 for π. Write the correct answer.

1. The Feathered Serpent Pyramid in Teotihuacan, Mexico is the third largest in the city. Its base is a square that measures 65 m on each side. The pyramid is 19.4 m high. What is the volume of the Feathered Serpent Pyramid?

   ____

2. The Sun Pyramid in Teotihuacan, Mexico, is larger than the Feathered Serpent Pyramid. The sides of the square base and the height are each about 3.3 times larger than the Feathered Serpent Pyramid. How many times larger is the volume of the Sun Pyramid than the Feathered Serpent Pyramid?

   ____

3. An oil funnel is in the shape of a cone. It has a diameter of 4 inches and a height of 6 inches. If the end of the funnel is plugged, how much oil can the funnel hold before it overflows?

   ____

4. One quart of oil has a volume of approximately 57.6 in^3. Does the oil funnel in exercise 3 hold more or less than 1 quart of oil?

   ____

Round to the nearest tenth. Use 3.14 for π. Choose the letter for the best answer.

5. An ice cream cone has a diameter of 4.2 cm and a height of 11.5 cm. What is the volume of the cone?

   A 18.7 cm^3
   B 25.3 cm^3
   C 53.1 cm^3
   D 212.3 cm^3

6. When decorating a cake, the frosting is put into a cone shaped bag and then squeezed out a hole at the tip of the cone. How much frosting is in a bag that has a radius of 1.5 inches and a height of 8.5 inches?

   F 5.0 in^3 **H** 15.2 in^3
   G 13.3 in^3 **J** 20.0 in^3

7. What is the volume of the hourglass at the right?

   A 13.1 in^3
   B 26.2 in^3
   C 52.3 in^3
   D 102.8 in^3

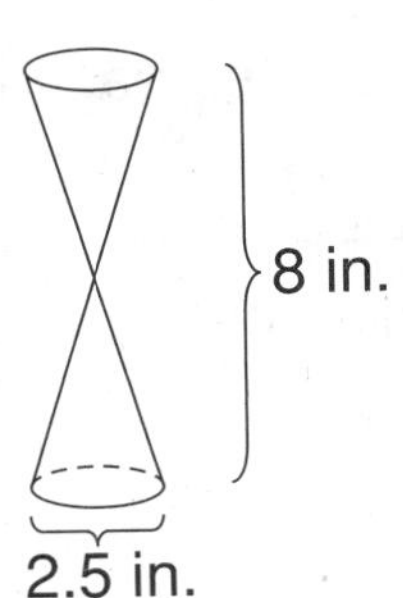

Name ______________________ Date __________ Class __________

LESSON 8-6

Reading Strategies

Compare and Contrast

Compare the shapes of these three-dimensional figures.

Cone

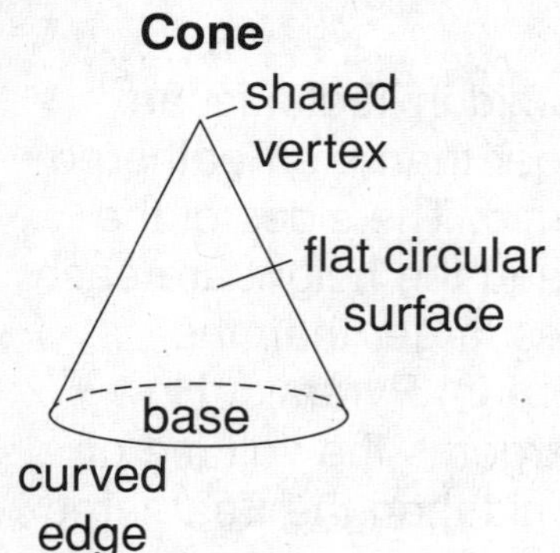

Pyramid

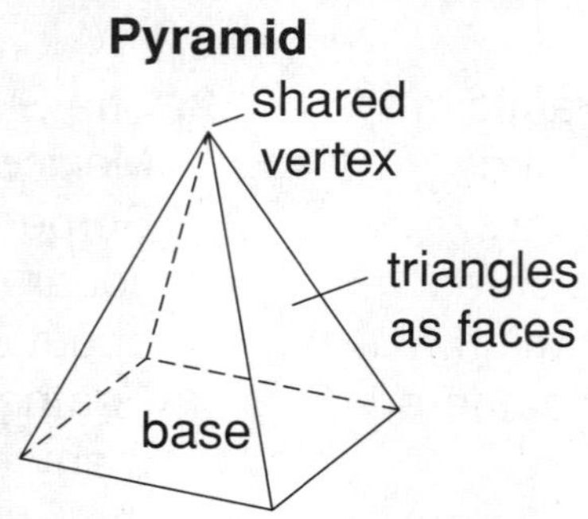

Compare the facts for the cone and the pyramid.

- three-dimensional figure
- base is a circle
- other face is a curved surface
- one vertex

- three-dimensional figure
- base is a polygon
- other faces are triangles
- at least four vertices

Compare the pictures and the facts about the cone and the pyramid to answer the questions.

1. Which figure has a circular base?

____

2. What is the shape of the base of a pyramid?

____

3. Which figure has triangles as faces?

____

4. How many bases does each figure have?

____

5. How many dimensions does each figure have?

____

6. How are the figures different?

____

Name ______________________ Date __________ Class __________

LESSON 8-6 Puzzles, Twisters & Teasers

How Much Volume!

Decide whether or not the volume given for each figure is correct. Circle the letter above your answer. Use the letters to solve the riddle.

1. $V = 21.8$

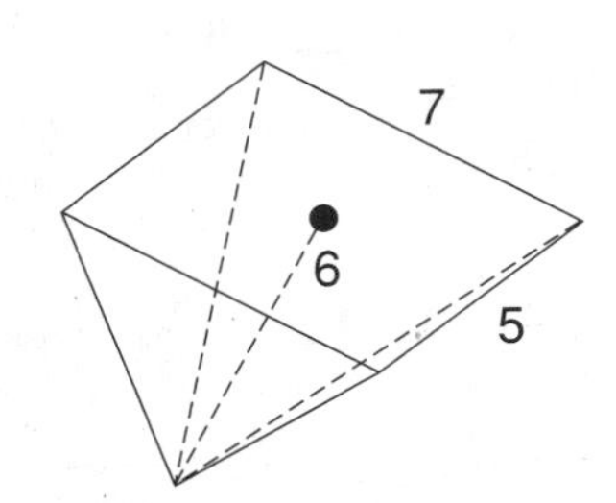

2. $V = 52.5$

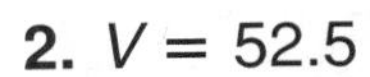

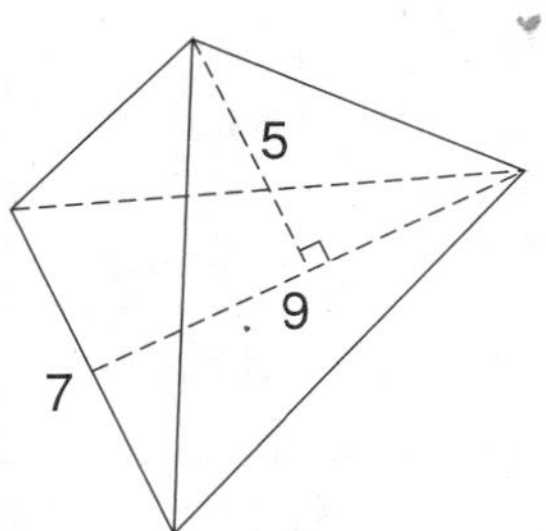

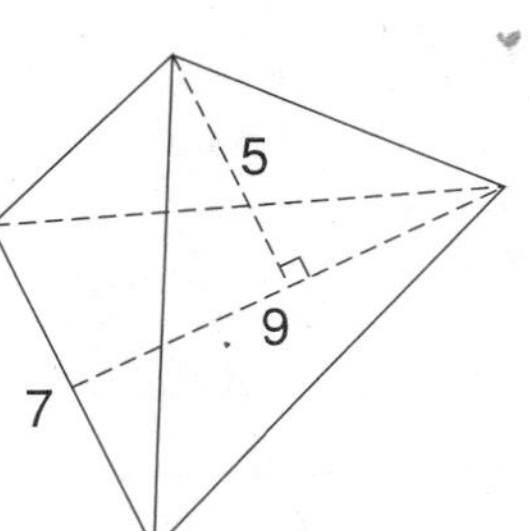

3. $V = 15$

5
3

D	B	R	E	L	A
correct	incorrect	correct	incorrect	correct	incorrect

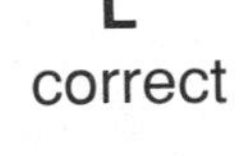

4. $V = 924$

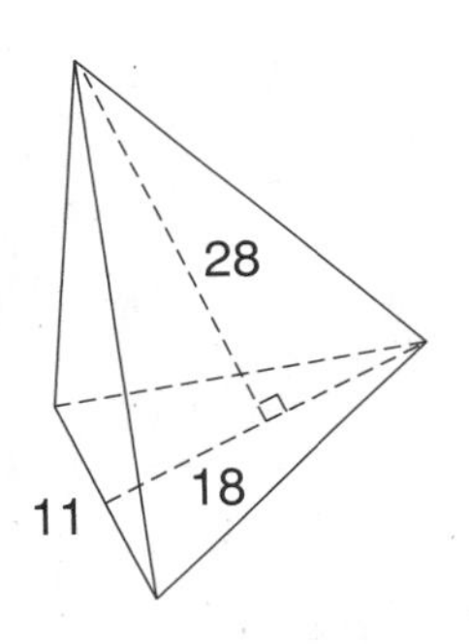

5. $V = 65.56$

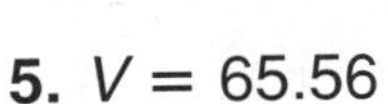

6.5
5.6

6. $V = 3159$

13
27
27

Z	R	K	I	L	T
correct	incorrect	correct	incorrect	correct	incorrect

What kind of nut is like a country?

A ____ ____ ____ ____ ____ ____ NUT
1 2 3 4 5 6

Name ______________________ Date ___________ Class ___________

LESSON 8-7

Practice A

Surface Area of Prisms and Cylinders

Find the surface area of each figure to the nearest tenth.
Prism: $S = 2B + Ph$. Cylinder: $S = 2\pi r^2 + 2\pi rh$.
Use 3.14 for π.

1.

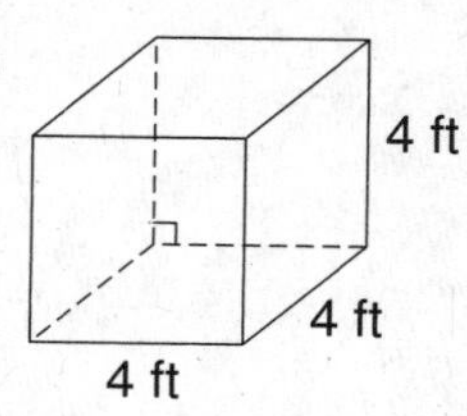

____

2.

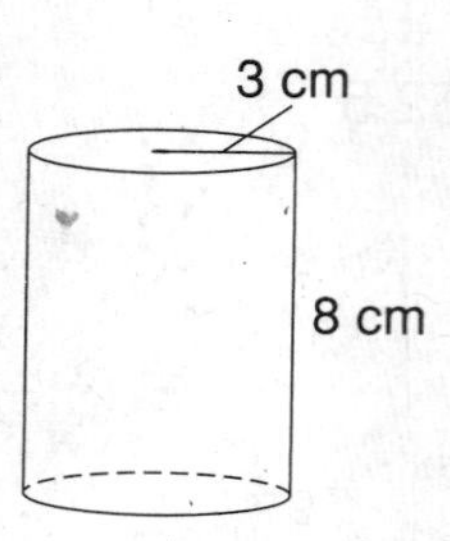

____

3.

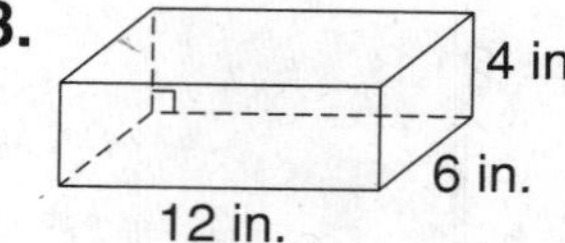

____

4.

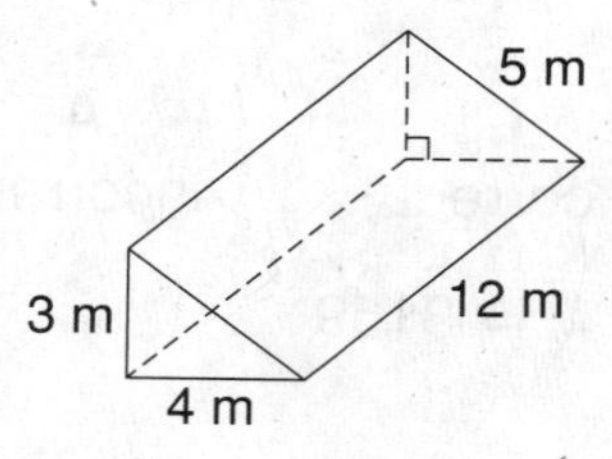

____

5.

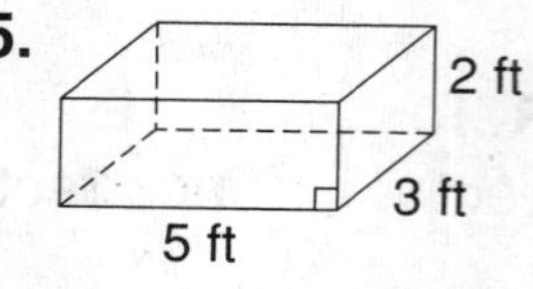

____

6.

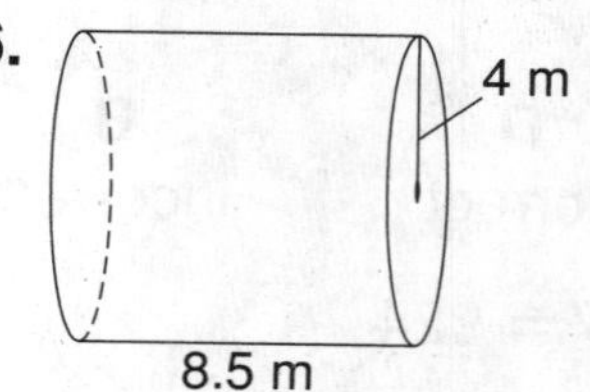

____

7.

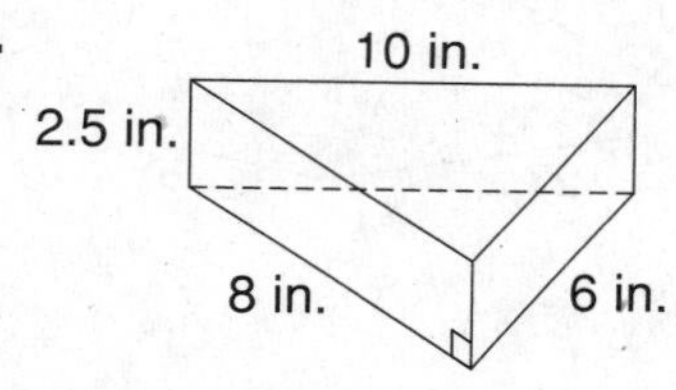

____

8.

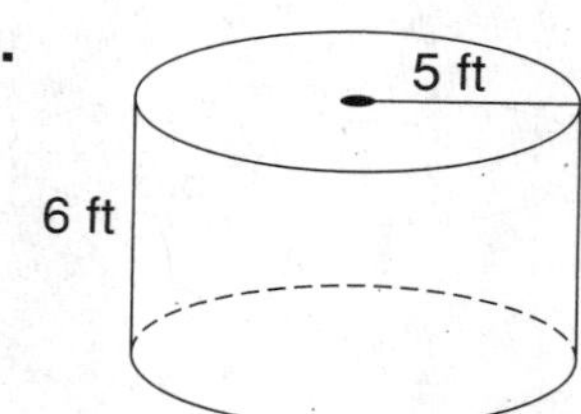

____

9.

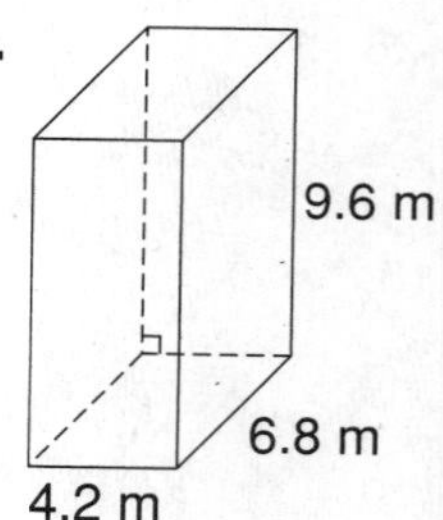

____

10. Find the surface area to the nearest tenth of a cylinder 50.6 ft tall that has a diameter of 30 ft. Use 3.14 for π. ________________

11. Find the surface area to the nearest tenth of a square prism with sides 6.2 m. ________________

12. To the nearest tenth, how much paper is needed for the label of a soup can if the can is 8.5 in. tall and has a diameter of 4 in.? Use 3.14 for π. (Hint: The label does not cover the bases of the cylinder.) ________________

Name ______________________ Date ____________ Class ____________

LESSON **8-7**

Practice B

Surface Area of Prisms and Cylinders

Find the surface area of each figure to the nearest tenth. Use 3.14 for π.

1.

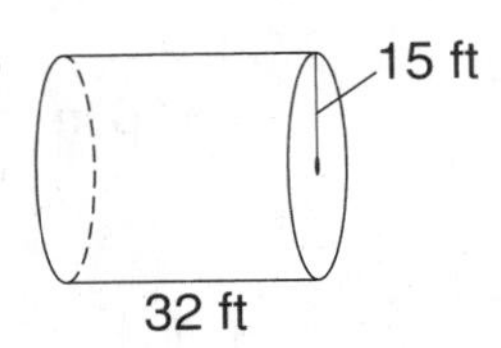

____

2.

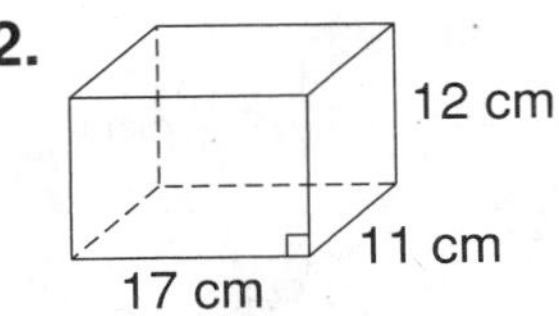

____

3.

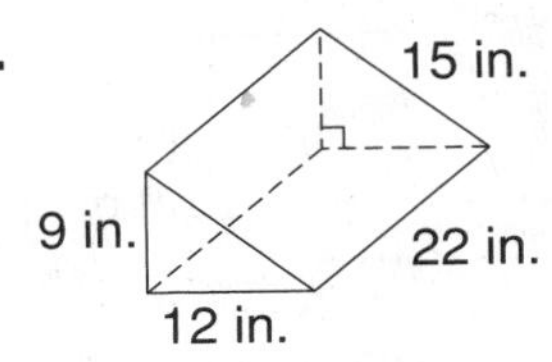

____

4.

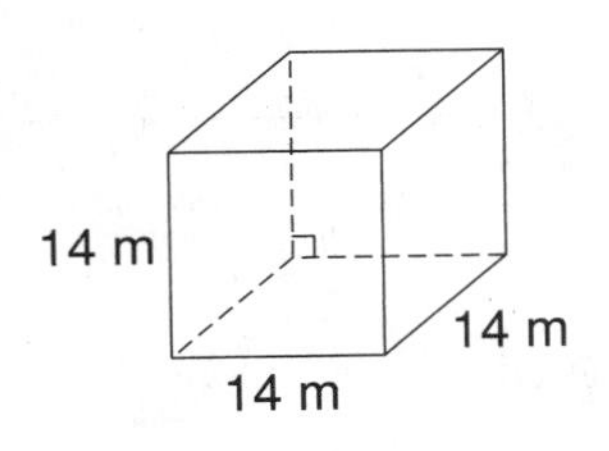

____

5.

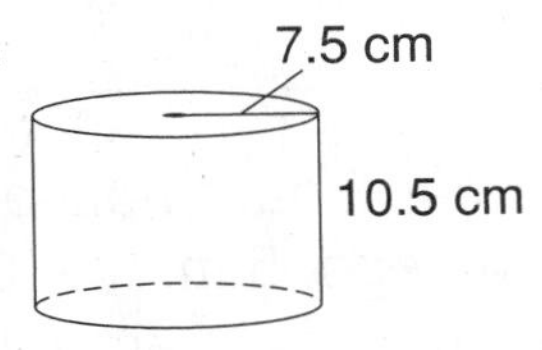

____

6.

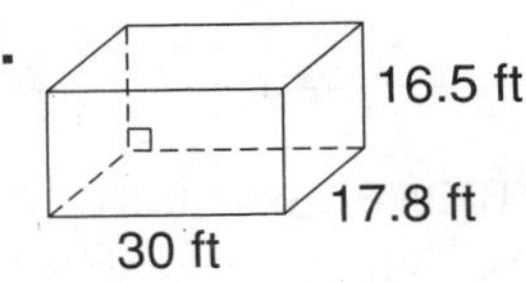

____

7.

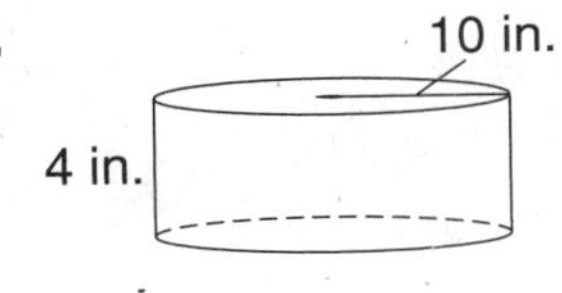

____

8.

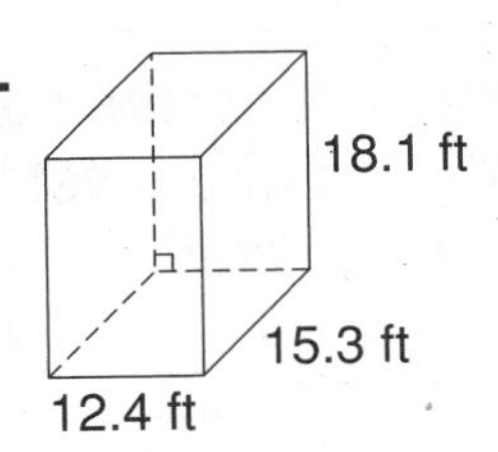

____

9.

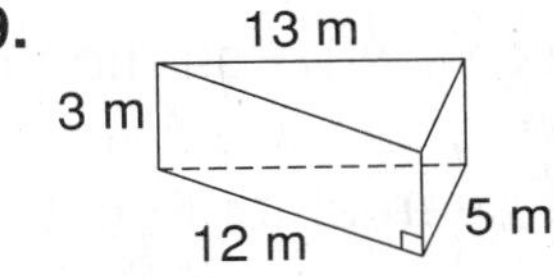

____

10. Find the surface area to the nearest tenth of a rectangular prism with height 15 m and sides 14 m and 13 m. ______________

11. Find the surface area to the nearest tenth of a cylinder 61.7 ft tall that has a diameter of 38 ft. ______________

12. Henry wants to paint the ceiling and walls of his living room. One gallon of paint covers 450 ft^2. The room is 24 ft by 18 ft, and the walls are 9 ft high. How many full gallons of paint will Henry need to paint his living room? ______________

13. A rectangular prism is 18 in. by 16 in. by 10 in. Explain the effect, if any, tripling all the dimensions will have on the surface area of the figure.

____

____

Name ______________________ Date __________ Class __________

LESSON 8-7

Practice C

Surface Area of Prisms and Cylinders

Find the surface area of each figure to the nearest tenth. Use 3.14 for π.

1.

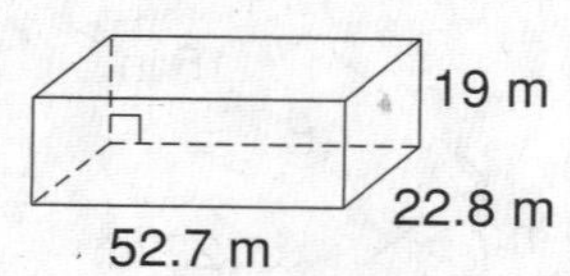

2.

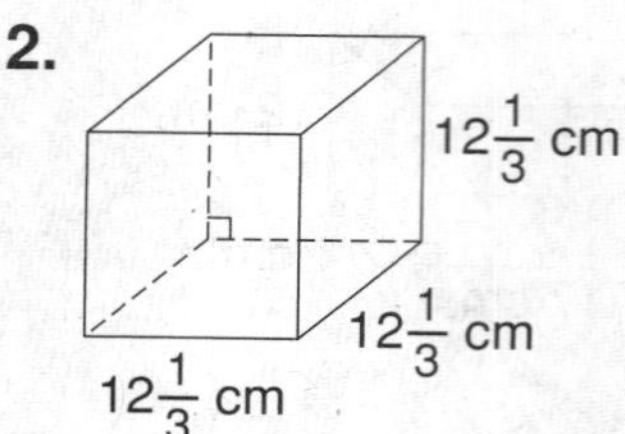

3.

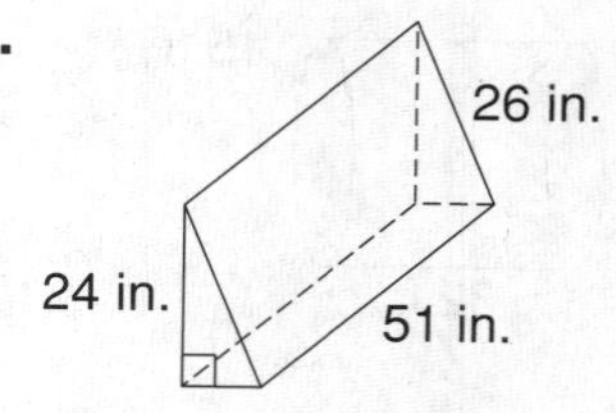

______________ ______________ ______________

4. Germane is making a rectangular cardboard house for his pet gerbil. He wants the house to be at least 5 in. tall, 6 in. wide, and 8 in. long. What is the minimum surface area, if the base of the rectangular prism is open?

____

5. Galena is a common isometric mineral. It occurs naturally in the form of a cube. What is the surface area of a crystal of galena to the nearest tenth if its sides measure 8.4 cm?

____

6. Cristina is painting the playhouse she made for her sister, Delia. She intends to paint only the outside of the playhouse, which does not have a floor. The paint she chose must be purchased in full gallons. Each gallon covers 200 ft^2 and costs $15.95 a gallon. How much will it cost Cristina to paint the playhouse?

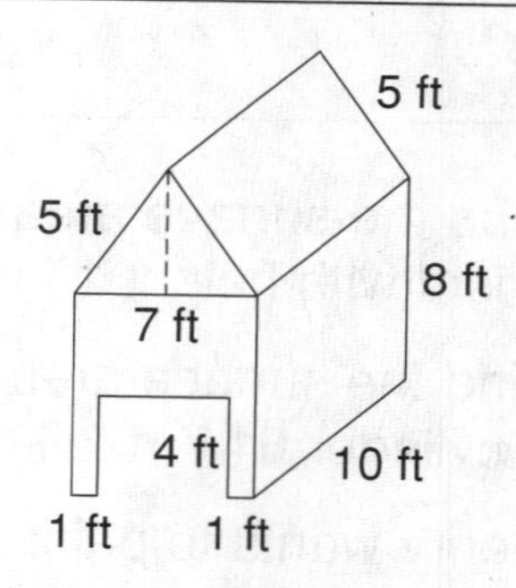

____

Find the missing dimension in each figure with the given surface area.

7. surface area = 1248 in^2

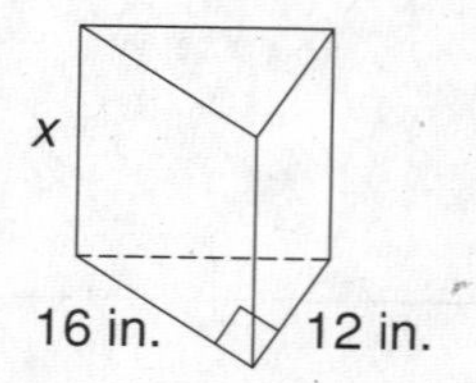

8. surface area = 628 ft^2

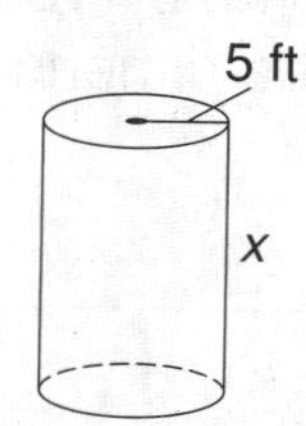

______________ ______________

Name ______________________ Date ____________ Class ____________

LESSON **8-7**

Reteach

Surface Area of Prisms and Cylinders

Find the number of tiles needed to cover the faces of the prism. Unfold the prism to get a better look at its six faces.

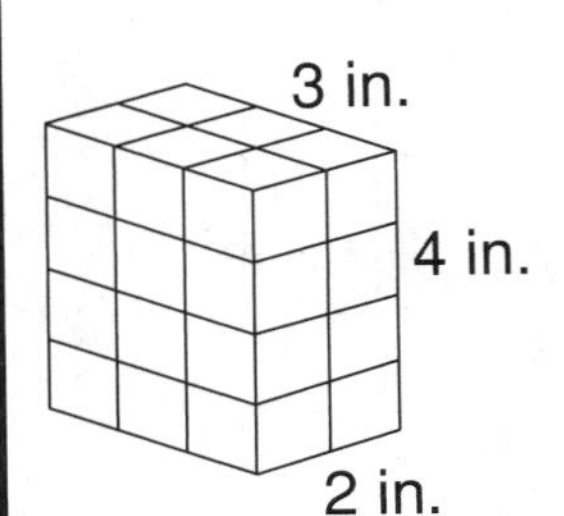

Face	Area	in^2
top	3×2	6
bottom	3×2	6
front	3×4	12
back	3×4	12
left	2×4	8
right	2×4	8
	total =	52 tiles

Surface Area S = the sum of the areas of the faces of the prism
= top + bottom + front + back + left + right
= area of bases + area of lateral faces
= $2B$ + perimeter of the base × height of prism

$\mathbf{S = 2B + Ph}$

$S = 2(3 \times 2) + (3 + 2 + 3 + 2) \times 4$

$S = 12 + (10) \times 4$

$S = 12 + 40$

$S = 52 \text{ in}^2$

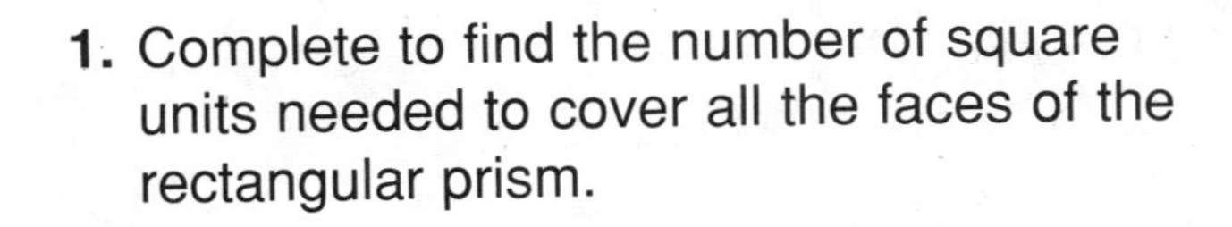

1. Complete to find the number of square units needed to cover all the faces of the rectangular prism.

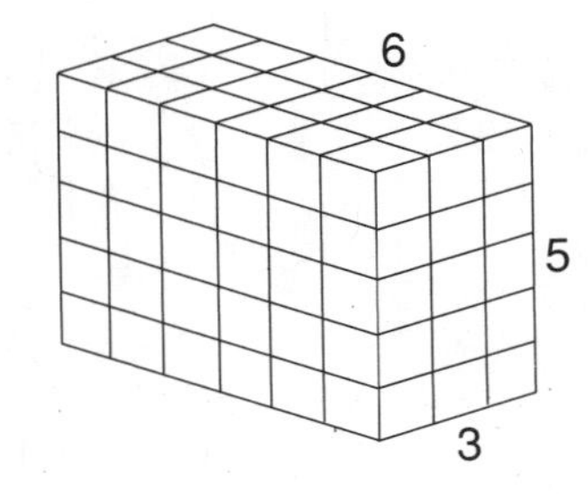

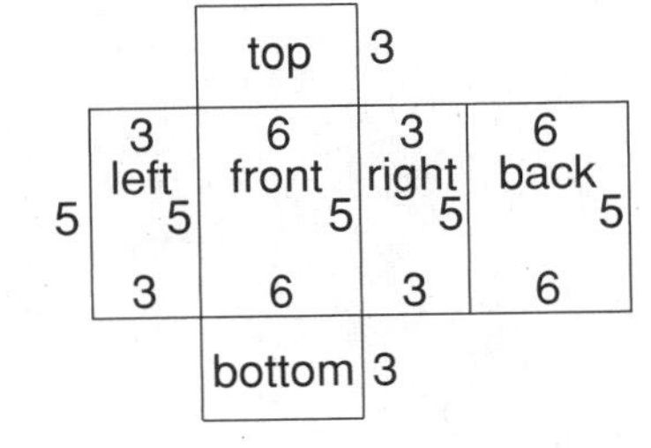

Face	Area	in^2
top	______	______
bottom	______	______
front	______	______
back	______	______
left	______	______
right	______	______
	total =	______

2. Complete to find the surface area of the prism.

$S = 2B + Ph$

$S = 2(\_\_\_\_\_\_) + (\_\_\_\_\_\_\_\_\_\_\_\_\_\_\_\_) \times \_\_\_\_$

$S = 2(\_\_\_\_\_\_) + (\_\_\_\_\_\_\_\_\_\_\_\_\_\_\_\_) \times \_\_\_\_$

$S = \_\_\_\_\_\_ + \_\_\_\_\_\_\_\_\_\_\_\_\_\_\_\_ = \_\_\_\_\_\_ \text{ in}^2$

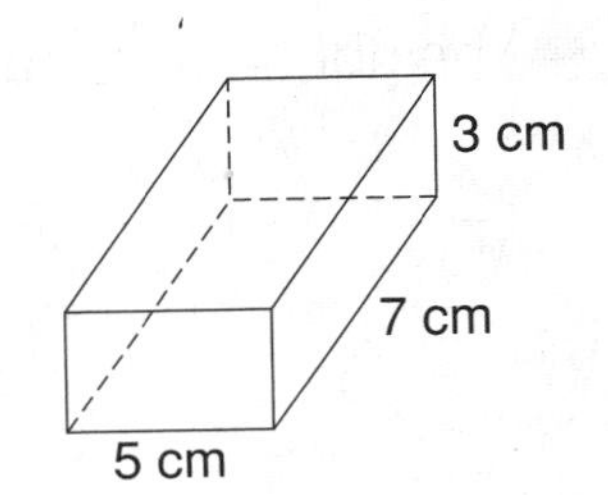

Name ______________________ Date __________ Class __________

LESSON 8-7

Reteach

Surface Area of Prisms and Cylinders (continued)

An unfolded cylinder results in two circles and a lateral surface drawn as a rectangle.

The base of the rectangle equals the circumference of the circular base.

The height of the rectangle equals the height of the cylinder.

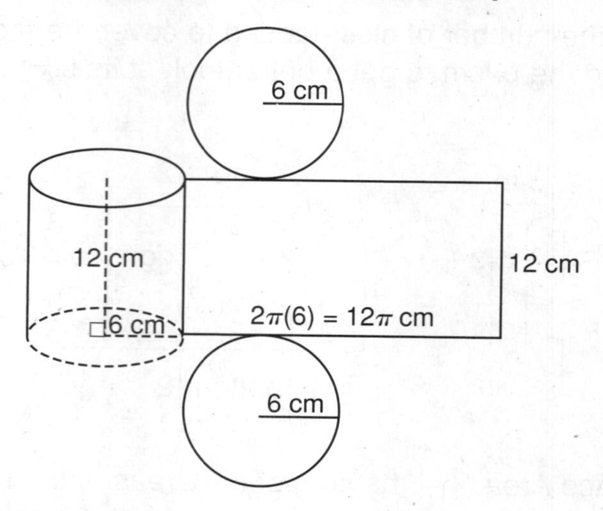

Surface Area S = area of 2 circular bases + area of lateral surface (rectangle)

$= 2(\pi r^2) +$ circumference × height

$= 2\pi r^2 + 2\pi r \times h$

$\mathbf{S = 2\pi r^2 + 2\pi rh}$

$S = 2\pi(6^2) + 2\pi(6)(12)$

$S = 72\pi + 144\pi = 216\pi \text{ cm}^2$

$S \approx 216(3.14) \approx 678.24 \text{ cm}^2$

3. Complete to find the surface area of the cylinder.

$\mathbf{S = 2\pi r^2 + 2\pi rh}$

$S = 2\pi($________$) + 2\pi \times$ ______ $\times$ ______

$S =$ ________ $\pi +$ ________ π

$S =$ ________ π

$S \approx$ ________ (3.14)

$S \approx$ ________ in^2

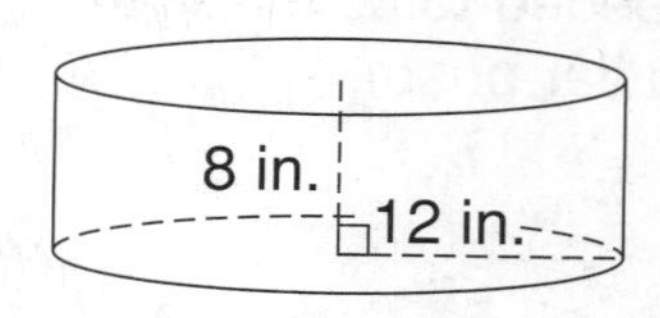

Find the surface area of each cylinder. Round to the nearest whole number.

4. height = 10 ft, radius = 5 ft

____

5. height = 2.5 cm, diameter = 8 cm

____

Name ______________________ Date __________ Class __________

Challenge

LESSON **8-7**

Eight Snips

A **cube** is a prism with six congruent square faces and eight vertices.

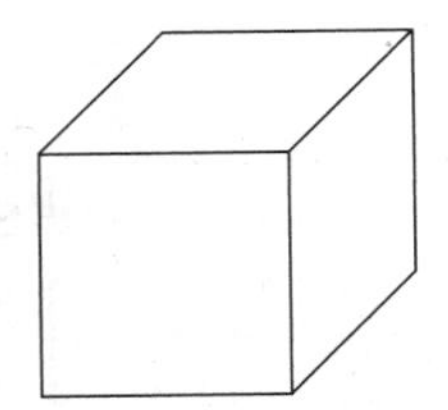

By cutting off the corners of the cube $\frac{1}{3}$ of the way into each edge, a **truncated cube** is created.

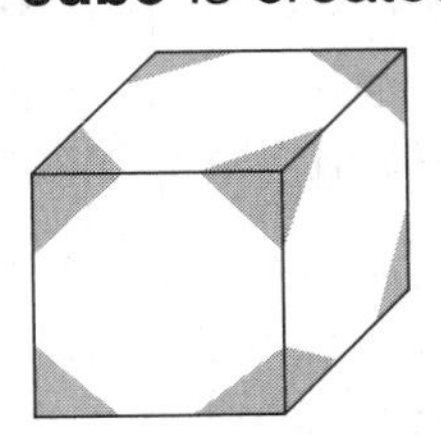

1. Cut along the outer perimeter of this pattern for a truncated cube. Carefully cut into the notched areas found next to some of the shaded sections.

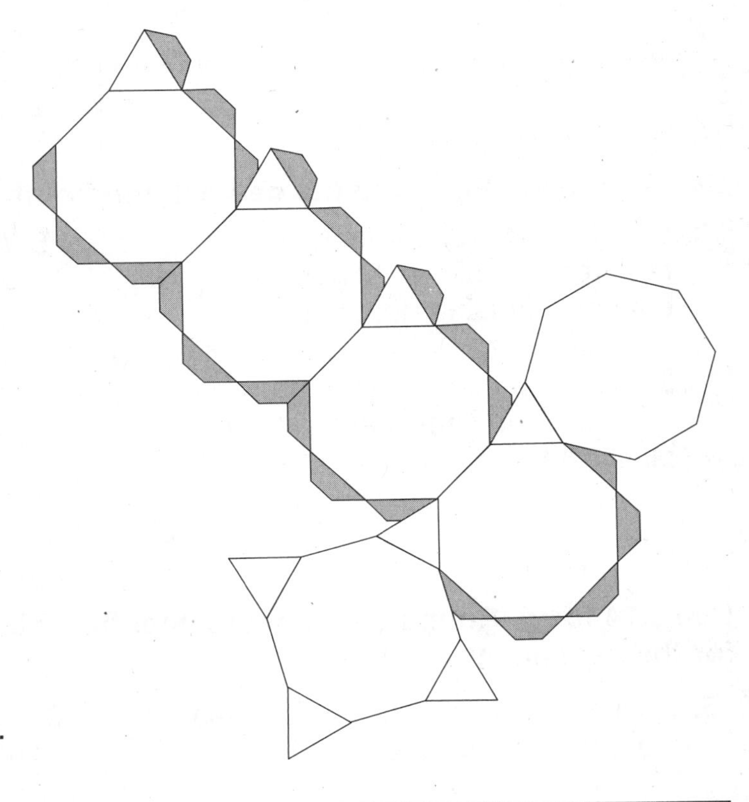

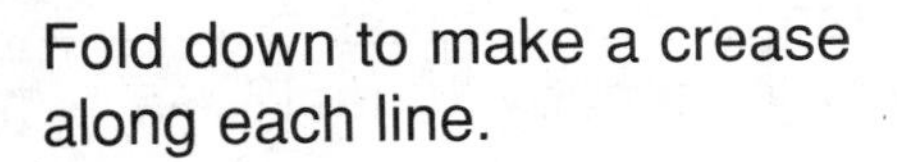

Fold down to make a crease along each line.

Using the shaded sections as "underlaps," tape the figure together.

2. How many faces in all?

____

3. Describe the nature of the faces.

____

4. Find the surface area of a truncated cube. The original cube had edge length = 24 in. Use $\sqrt{3} \approx 1.73$. Answer to the nearest tenth of a square inch.

____

Name ______________________ Date ____________ Class ____________

LESSON **8-7**

Problem Solving

Surface Area of Prisms and Cylinders

An important factor in designing packaging for a product is the amount of material required to make the package. Consider the three figures described in the table below. Use 3.14 for π. Round to the nearest tenth. Write the correct answer.

1. Find the surface area of each package given in the table.

2. Which package has the lowest materials cost? Assume all of the packages are made from the same material.

____

Package	Dimensions	Volume	Surface Area
Prism	Base: 2" × 16" Height = 2"	64 in^3	
Prism	Base: 4" × 4" Height = 4"	64 in^3	
Cylinder	Radius = 2" Height = 5.1"	64.06 in^3	

Use 3.14 for π. Round to the nearest hundredth.

3. How much cardboard material is required to make a cylindrical oatmeal container that has a diameter of 12.5 cm and a height of 24 cm, assuming there is no overlap? The container will have a plastic lid.

____

4. What is the surface area of a rectangular prism that is 5 feet by 6 feet by 10 feet?

____

Use 3.14 for π. Round to the nearest tenth. Choose the letter for the best answer.

5. How much metal is required to make the trough pictured below?

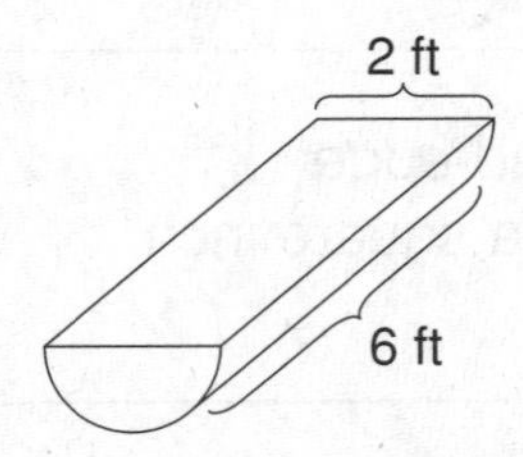

A 22.0 ft^2
B 34.0 ft^2
C 44.0 ft^2
D 56.7 ft^2

6. A can of vegetables has a diameter of 9.8 cm and is 13.2 cm tall. How much paper is required to make the label, assuming there is no overlap? Round to the nearest tenth.

F 203.1 cm^2
G 406.2 cm^2
H 557.0 cm^2
J 812.4 cm^2

Name ______________________ Date __________ Class __________

LESSON 8-7

Reading Strategies

Identify Relationships

The **surface area** of a solid figure is the total area of its outside surfaces. You can think of surface area as the part of the solid shape that you can paint. You can paint all the surfaces. Surface area is always given in square units, just like area.

In order to figure out the surface area of a rectangular prism, you can "unfold" it to make a net.

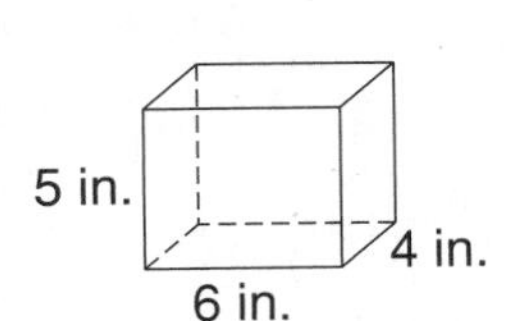

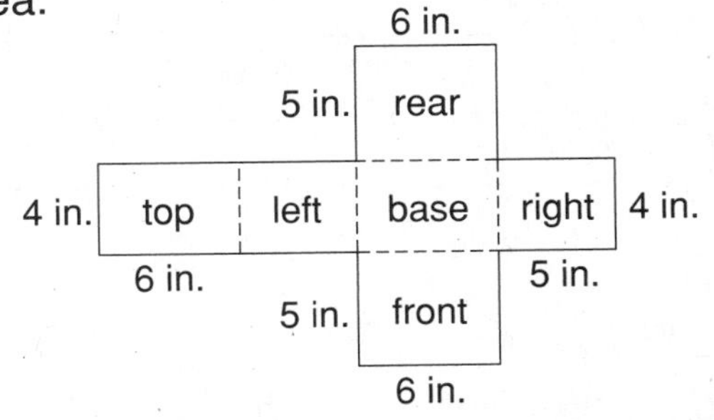

A rectangular prism has six faces, each the shape of a rectangle. To find the surface area of a rectangular prism, find the sum of the areas of the six faces, or rectangles. Opposite faces are equal.

The figure above will help you find the surface area of the rectangular prism. Use the formula $A = (\ell \cdot w)$ to find the area of each face.

Answer each question to find the surface area of the rectangular prism.

1. What is the area of the base rectangle? ______________________
2. What is the area of the top rectangle? ______________________
3. What is the area of the rear rectangle? ______________________
4. What is the area of the front rectangle? ______________________
5. What is the area of the right rectangle? ______________________
6. What is the area of the left rectangle? ______________________

To find the surface area of the rectangular prism, add the area of all six rectangles.

7. What is the surface area of the rectangular prism?

____

Name ______________________ Date __________ Class __________

LESSON 8-7

Puzzles, Twisters & Teasers

How Much Volume!

Decide which formula should be used to find the surface area of each figure. Circle the letter above your answer. Then use the letters to solve the riddle.

1.

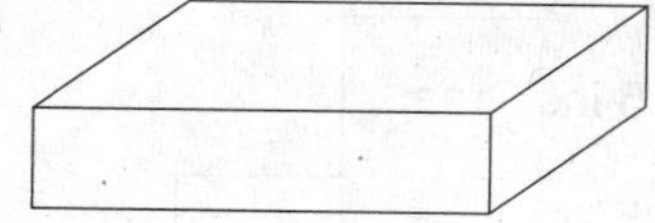

W $2B + ph$ R $2\pi r^2 + 2\pi rh$

2.

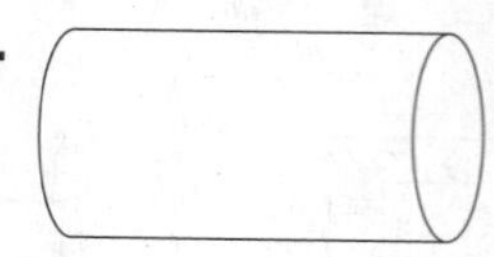

F $2B + ph$ T $2\pi r^2 + 2\pi rh$

3.

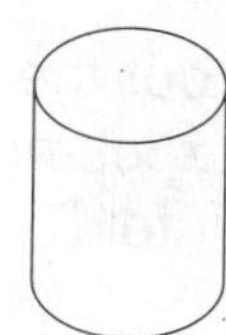

M $2B + ph$ R $2\pi r^2 + 2\pi rh$

4.

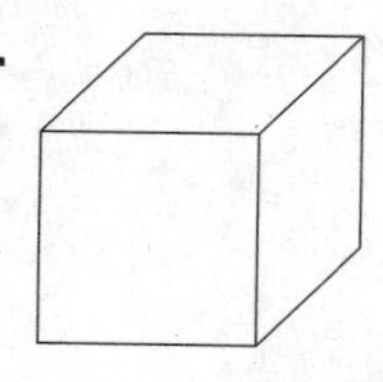

M $2B + ph$ G $2\pi r^2 + 2\pi rh$

5.

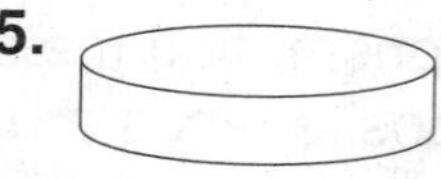

P $2B + ph$ L $2\pi r^2 + 2\pi rh$

6.

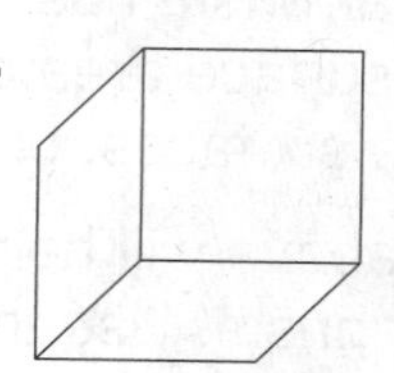

O $2B + ph$ Q $2\pi r^2 + 2\pi rh$

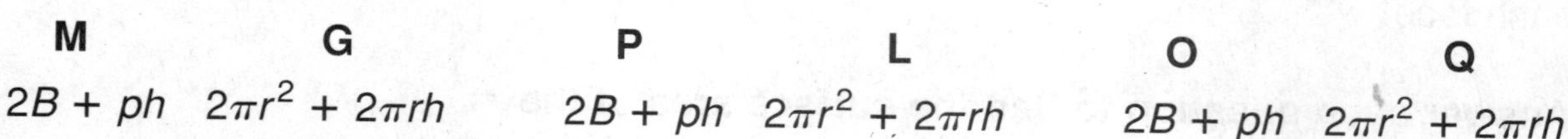

When would you go on red and stop on green?

When you're eating a ___ A ___ E ___ ___ E ___ ___ N.
(blanks numbered 1, 2, 3, 4, 5, 6)

Name ______________________ Date ____________ Class ____________

LESSON 8-8

Practice A

Surface Area of Pyramids and Cones

Find the surface area of each figure to the nearest tenth.
Pyramid: $S = B + \frac{1}{2}P\ell$. Cone: $S = \pi r^2 + \pi r\ell$.
Use 3.14 for π.

1.

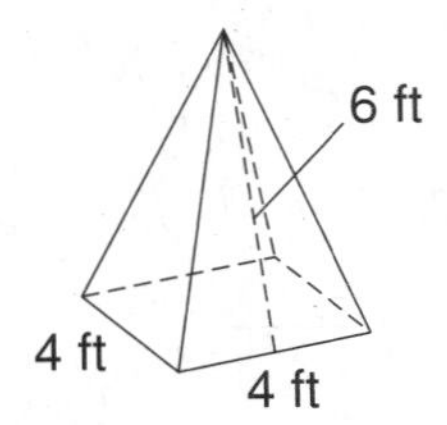

____

2.

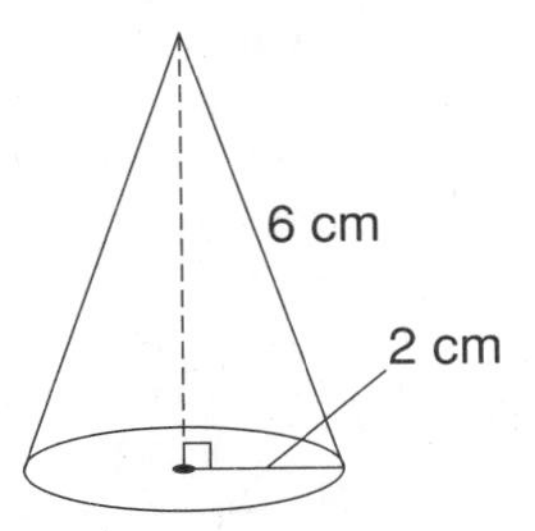

____

3.

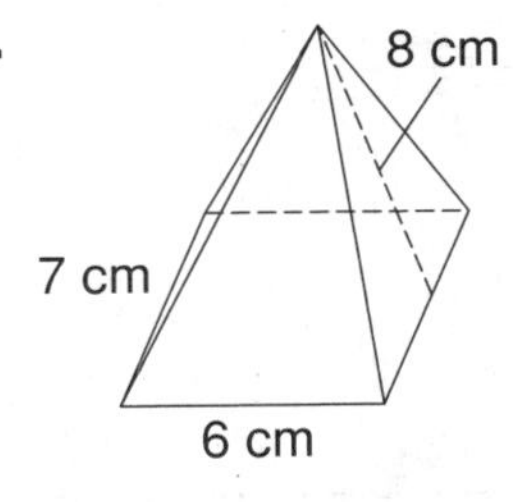

____

4.

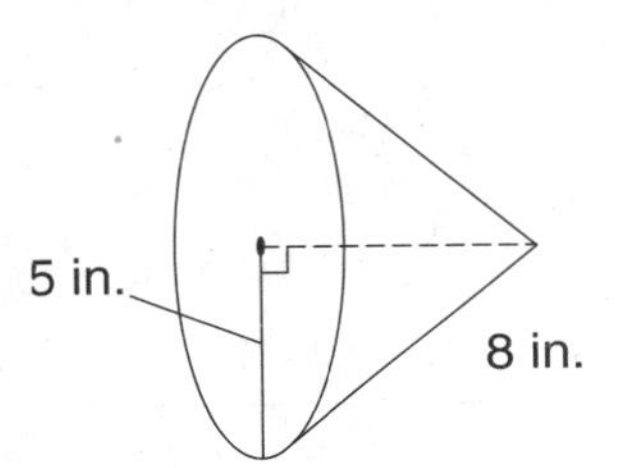

____

5.

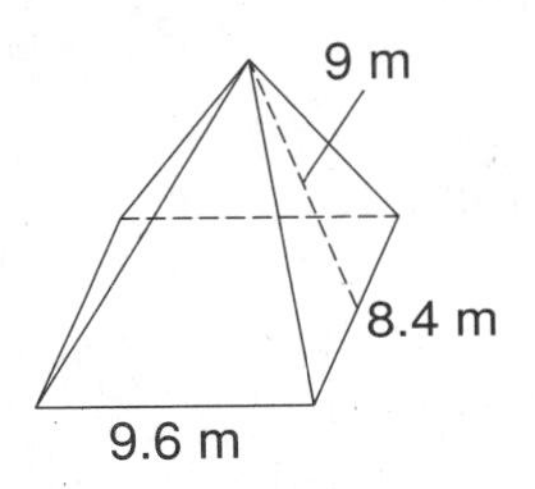

____

6.

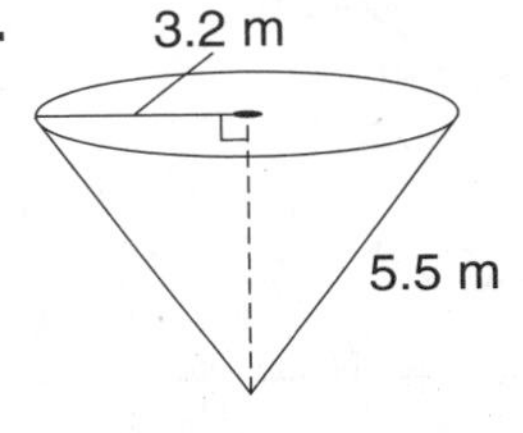

____

7.

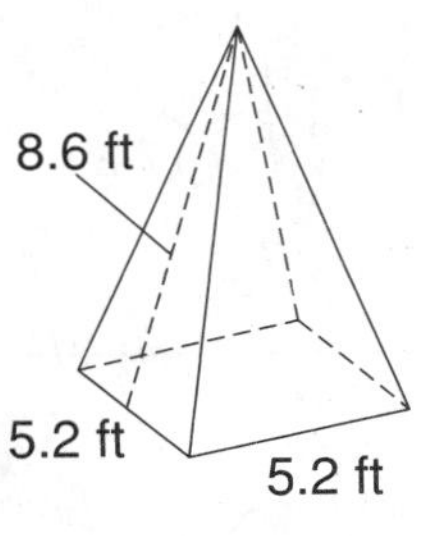

____

8.

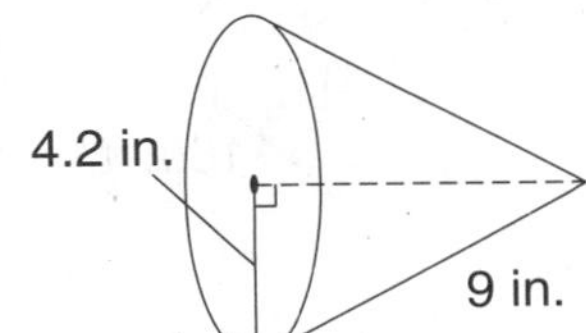

____

9.

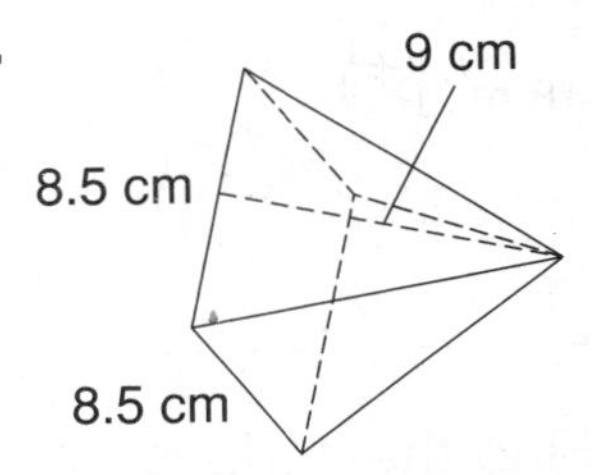

____

10. Find the surface area of a regular square pyramid with a slant height of 7 m and a base perimeter of 10 m.

____

11. Find the length of the slant height of a square pyramid if one side of the base is 5 ft and the surface area is 125 ft^2.

____

12. Find the length of the slant height of a cone with a radius of 5 cm and a surface area of 235.5 cm^2. Use 3.14 for π.

____

Name ______________________________ Date ____________ Class ____________

LESSON 8-8

Practice B

Surface Area of Pyramids and Cones

Find the surface area of each figure to the nearest tenth. Use 3.14 for π.

1.

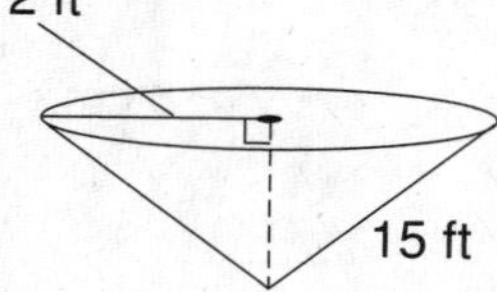

____

2.

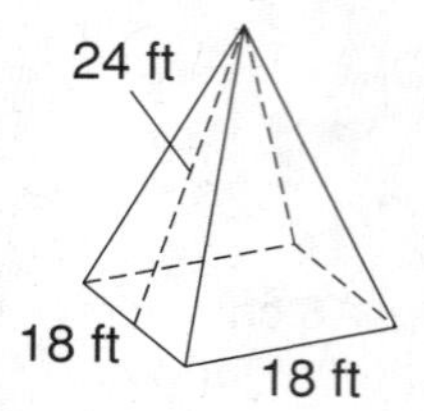

____

3.

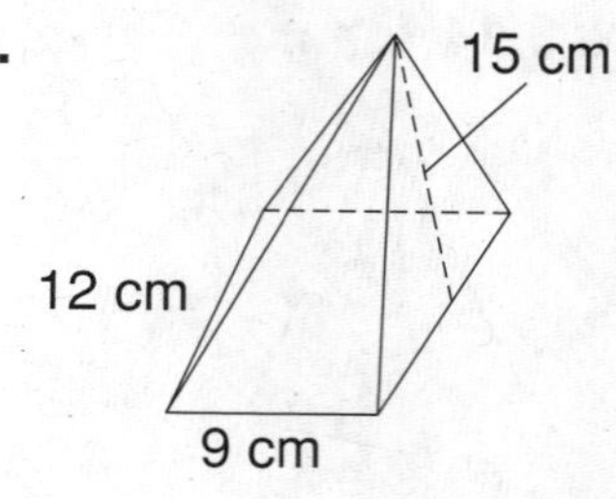

____

4.

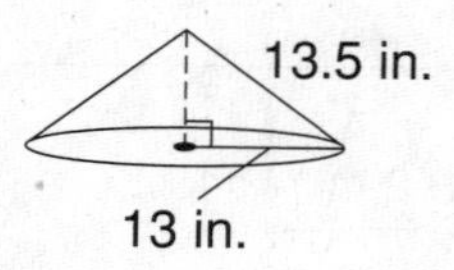

____

5.

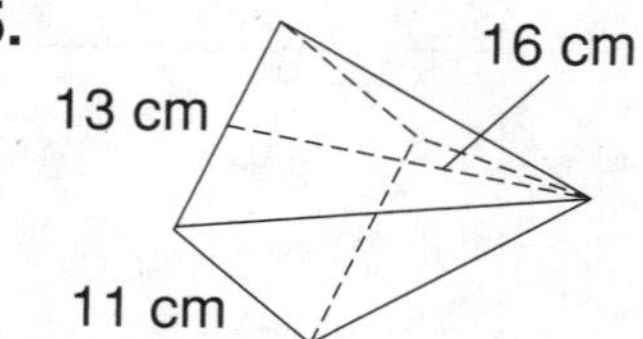

____

6.

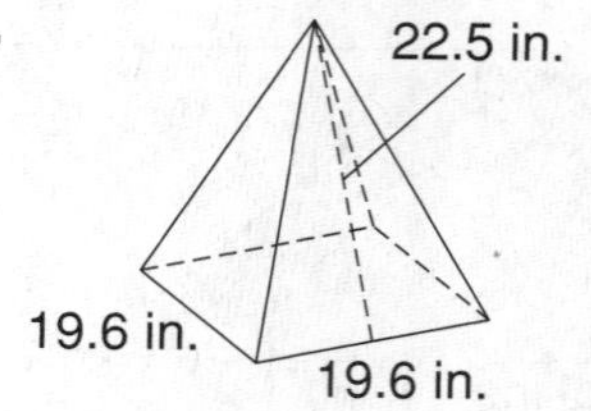

____

7.

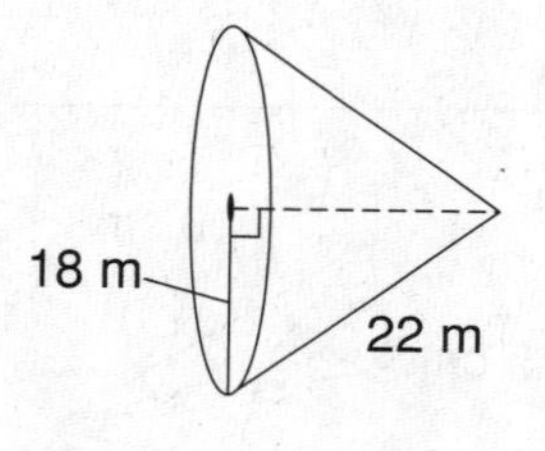

____

8.

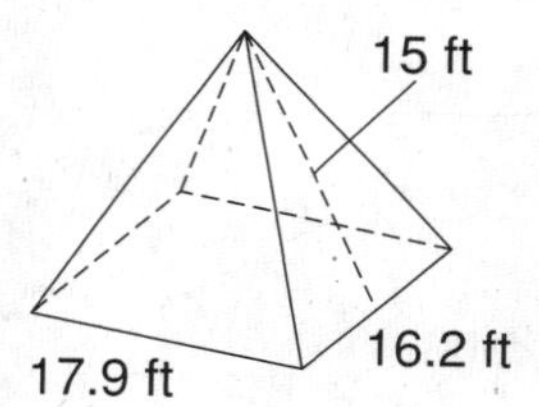

____

9.

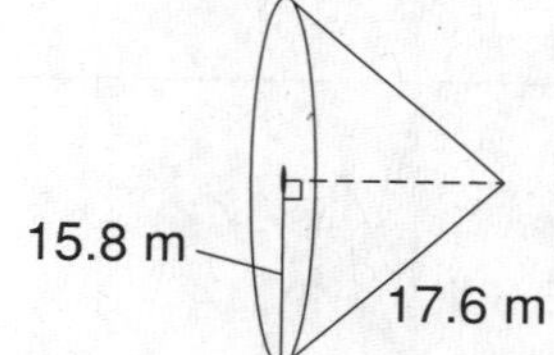

____

10. Find the surface area of a regular square pyramid with a slant height of 17 m and a base perimeter of 44 m. ______________

11. Find the length of the slant height of a square pyramid if one side of the base is 15 ft and the surface area is 765 ft^2. ______________

12. Find the length of the slant height of a cone with a radius of 15 cm and a surface area of 1884 cm^2. ______________

13. A cone has a diameter of 12 ft and a slant height of 20 ft. Explain whether tripling both dimensions would triple the surface area.

____

____

____

Name ______________________ Date __________ Class __________

LESSON 8-8

Practice C

Surface Area of Pyramids and Cones

Find the surface area of each figure to the nearest tenth. Use 3.14 for π.

1.

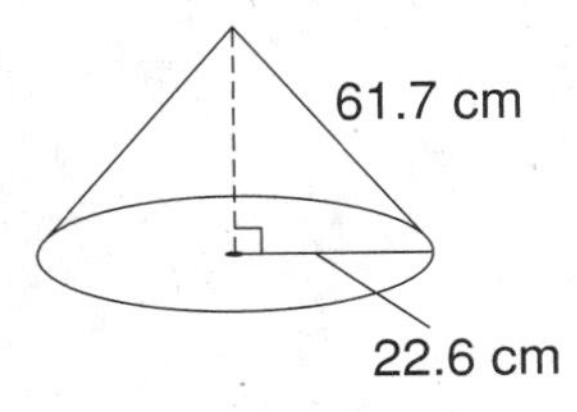

2.

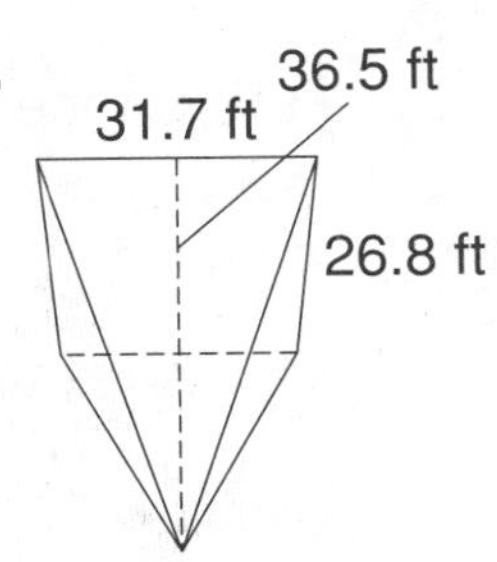

3.

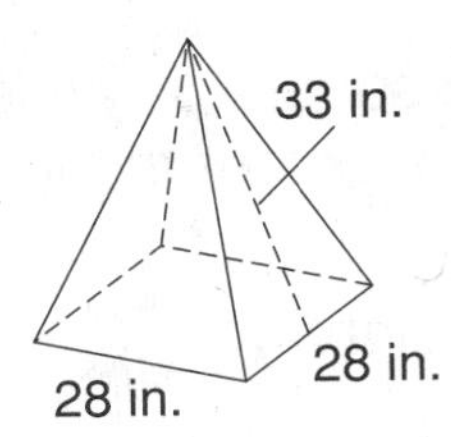

________ ________ ________

Find the surface area of each figure with the given dimensions. Use 3.14 for π.

4. regular square pyramid:
   base perimeter = 80 ft
   slant height = 25 ft

   ____

5. regular triangular pyramid:
   base area = 128 in^2
   base perimeter = 48 in.
   slant height = 21 in.

   ____

6. cone:
   diameter = 31 m
   slant height = 37 m

   ____

7. cone:
   radius = $5\frac{1}{3}$ cm
   slant height = 8.5 cm

   ____

8. A cone-shaped roof is attached to a silo. If the diameter of the roof is 25 ft and the slant height is 15 ft, find the surface area of the cone-shaped roof to the nearest tenth. ________

9. At 146.5 m high, the Great Pyramid of Egypt stood as the tallest structure in the world for more than 4000 years. It is believed that this great structure was built around 3200 B.C. One base side of the Great Pyramid is approximately 231 m and its slant height is approximately 186 m. Find the approximate surface area of the structure. ________

10. If a model of the Great Pyramid were built with a base side of 7.7 cm and slant height 6.2 cm, what would be the surface area of the model to the nearest tenth? ________

Name ______________________ Date __________ Class __________

LESSON 8-8

Reteach

Surface Area of Pyramids and Cones

Regular Pyramid:
base is a regular polygon; lateral faces are congruent triangles

When a square pyramid is unfolded, there are 5 faces: a square and 4 congruent triangles.

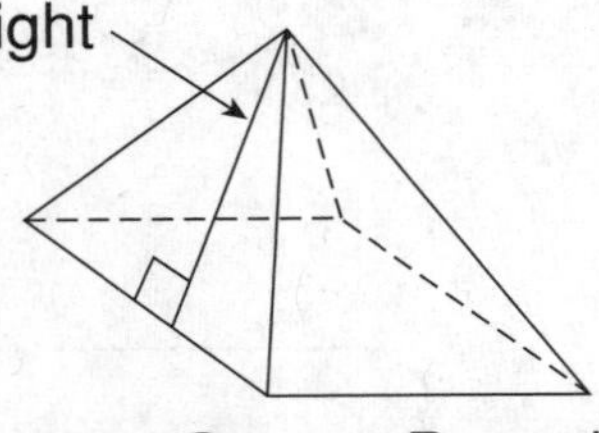

Square Pyramid

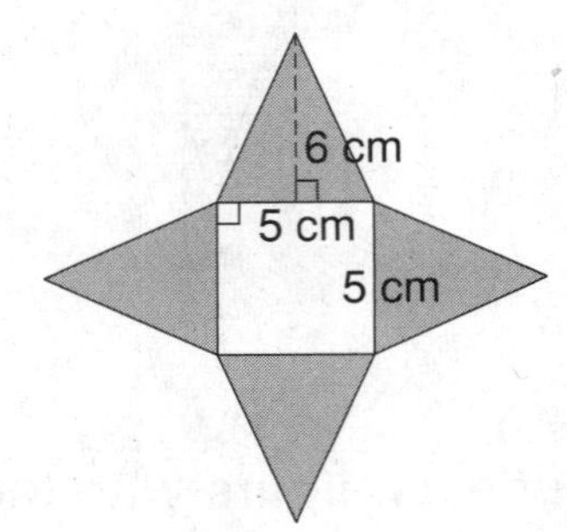

Surface Area S = the sum of the areas of the faces of the pyramid
= area of base + area of lateral faces

Surface Area S = area of square + 4(area of triangle)
$= B + \frac{1}{2}$ perimeter P of base × slant height ℓ of prism

$$\boldsymbol{S = B + \frac{1}{2}P\ell}$$

$S = (5 \times 5) + \frac{1}{2} \times (5 \times 4) \times 6$

$S = 25 + 60$

$S = 85\text{ cm}^2$

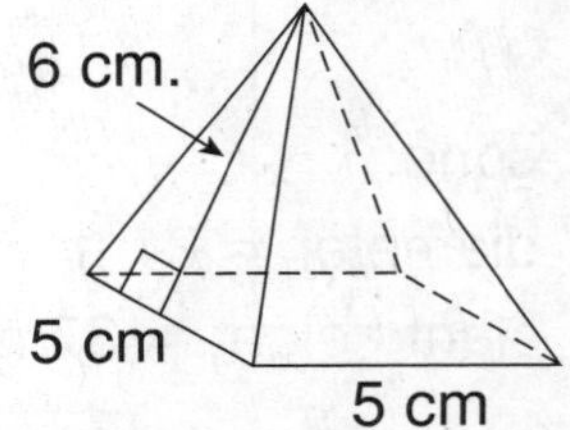

1. Find the surface area of the square pyramid.

   S = area of square + 4(area of triangle)

   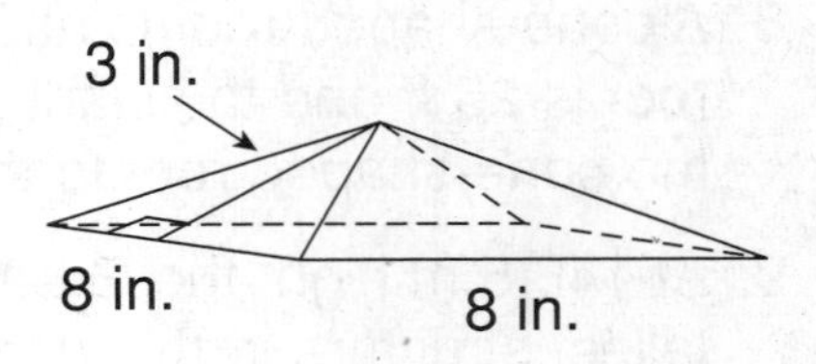

   S = __________ + __________

   S = __________ + __________

   S = ____________________ in^2

2. Complete to find the surface area of the square pyramid.

   $S = B + \frac{1}{2}P\ell$

   S = __________ $+ \frac{1}{2}$ __________ × __________

   S = __________ + __________

   S = ____________________ ft^2

   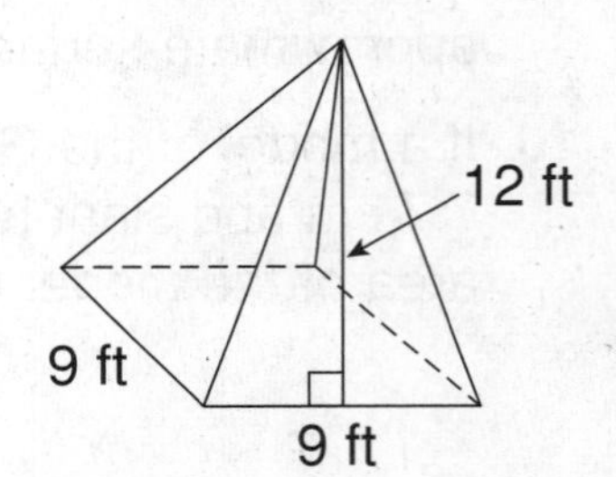

Name ______________________ Date __________ Class __________

LESSON 8-8

Reteach

Surface Area of Pyramids and Cones (continued)

An unfolded cone results in a circle and a lateral surface drawn as a sector of a circle.

The slant height of the cone is the radius of the circle sector.

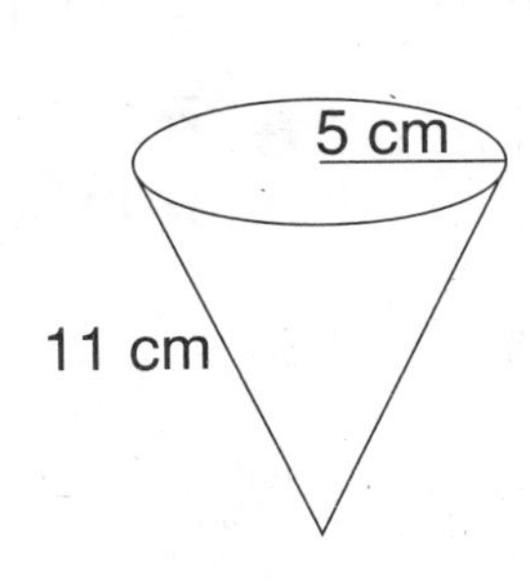

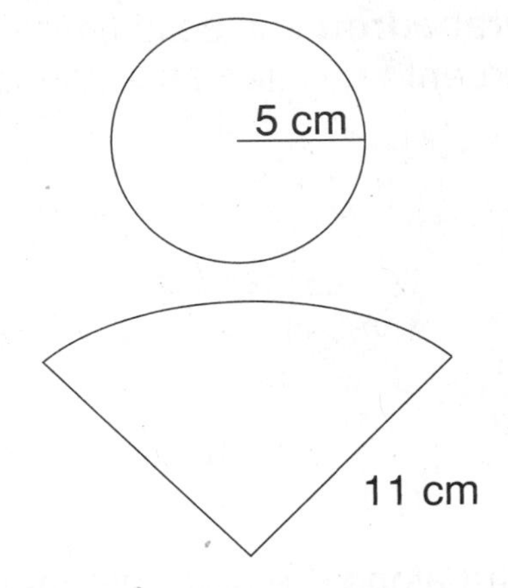

Surface Area S = area of circular base + area of lateral surface (circle sector)

$= \pi r^2 + \frac{1}{2}$ circumference of base × slant height

$= \pi r^2 + \frac{1}{2}(2\pi r) \times \ell$

$\boldsymbol{S = \pi r^2 + \pi r \ell}$

$S = \pi(5^2) + \pi(5)(11)$

$S = 25\pi + 55\pi = 80\pi \text{ cm}^2$

$S \approx 80(3.14) \approx 251.2 \text{ cm}^2$

Complete to find the surface area of the cone.

3. $S = \pi r^2 + \pi r \ell$

   $S = \pi($________$) + \pi \times$ ______ $\times$ ______

   $S =$ ________ $\pi +$ ________ $\pi =$ ________ π

   $S \approx$ ________ (3.14)

   $S \approx$ ____________ in^2

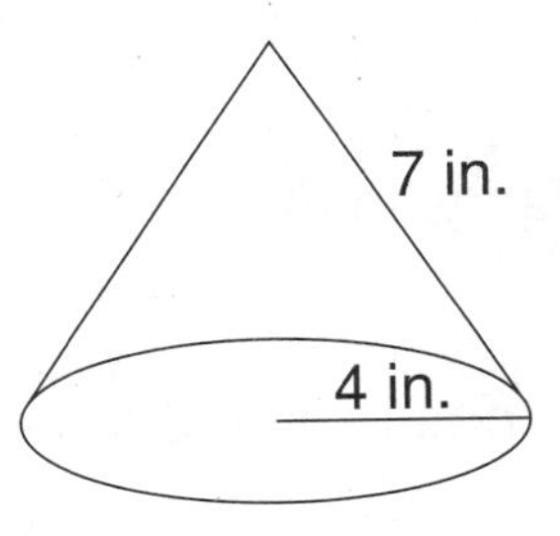

Find the surface area of each cone. Round to the nearest whole number.

4. radius = 3 ft, slant height = 5 ft

   ____

5. diameter = 8.6 cm, slant height = 10 cm

   ____

Name ______________________ Date ___________ Class ___________

LESSON 8-8

Challenge
Tongue Twister

A **tetrahedron** is a solid figure with four congruent equilateral triangle faces.

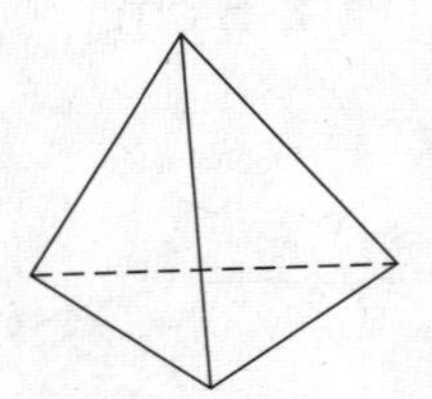

By cutting off the tips of the tetrahedron $\frac{1}{3}$ of the way into each edge, a **truncated tetrahedron** is created.

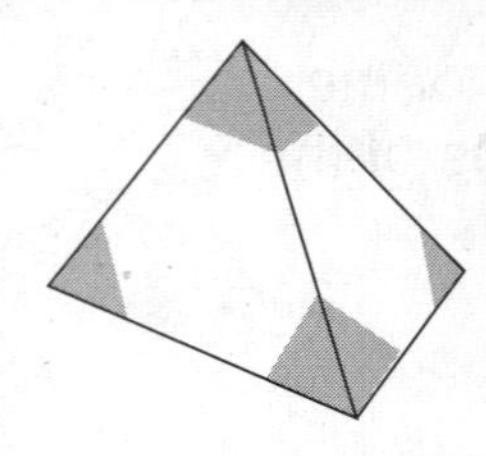

1. Cut along the outer perimeter of this pattern for a truncated tetrahedron. Carefully cut into the notched areas found next to some of the shaded sections.

   Fold down to make a crease along each line.

   Using the shaded sections as "underlaps," tape the figure together.

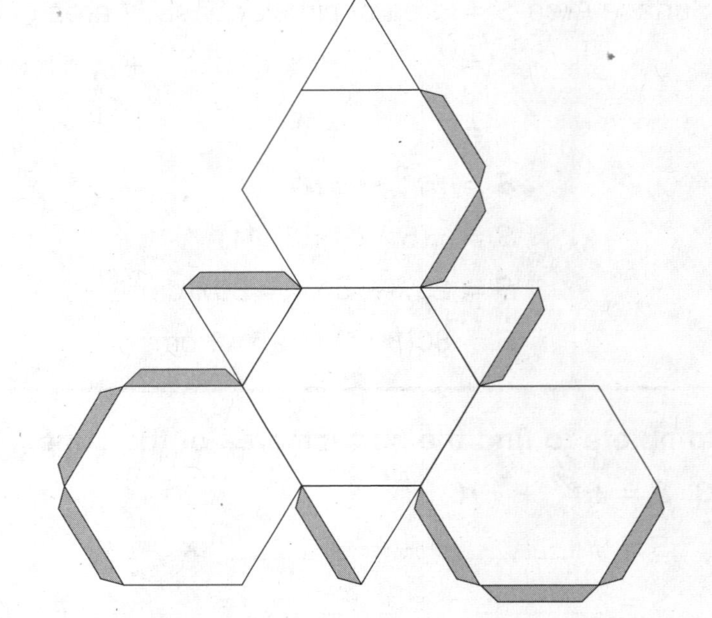

2. How many faces in all?

   ____

3. Describe the nature of the faces.

   ____

4. Area of Equilateral Triangle $= \frac{s^2}{4}\sqrt{3}$ Area of Regular Hexagon $= \frac{3s^2}{2}\sqrt{3}$

   Find the surface area of a truncated tetrahedron with $s = 2$ in. Answer in terms of $\sqrt{3}$. (*Hint:* Add like terms, such as $5\sqrt{3} + 2\sqrt{3} = 7\sqrt{3}$)

   ____

Name ______________________ Date ____________ Class ____________

LESSON **8-8**

Problem Solving

Surface Area of Pyramids and Cones

Round to the nearest tenth. Use 3.14 for π. Write the correct answer.

1. The Feathered Serpent Pyramid in Teotihuacan, Mexico, is the third largest in the city. Its base is a square that measures 65 m on each side. The pyramid is 19.4 m high and has a slant height of 37.8 m. The lateral faces of the pyramid are decorated with paintings. What is the surface area of the painted faces?

   ____

2. The Sun Pyramid in Teotihuacan, Mexico, is larger than the Feathered Serpent Pyramid. The sides of the square base and the slant height are each about 3.3 times larger than the Feathered Serpent Pyramid. How many times larger is the surface area of the lateral faces of the Sun Pyramid than the Feathered Serpent Pyramid?

   ____

3. An oil funnel is in the shape of a cone. It has a diameter of 4 inches and a slant height of 6 inches. How much material does it take to make a funnel with these dimensions?

   ____

4. If the diameter of the funnel in Exercise 6 is doubled, by how much does it increase the surface area of the funnel?

   ____

Round to the nearest tenth. Use 3.14 for π. Choose the letter for the best answer.

5. An ice cream cone has a diameter of 4.2 cm and a slant height of 11.5 cm. What is the surface area of the ice cream cone?

   A 4.7 cm^2 **C** 75.83 cm^2
   B 19.9 cm^2 **D** 159.2 cm^2

6. A marker has a conical tip. The diameter of the tip is 1 cm and the slant height is 0.7 cm. What is the area of the writing surface of the marker tip?

   F 1.1 cm^2 **H** 2.2 cm^2
   G 1.9 cm^2 **J** 5.3 cm^2

7. A skylight is shaped like a square pyramid. Each panel has a 4 m base. The slant height is 2 m, and the base is open. The installation cost is $5.25 per square meter. What is the cost to install 4 skylights?

   A $64 **C** $218
   B $159 **D** $336

8. A paper drinking cup shaped like a cone has a 10 cm slant height and an 8 cm diameter. What is the surface area of the cone?

   F 88.9 cm^2 **H** 251.2 cm^2
   G 125.6 cm^2 **J** 301.2 cm^2

Name ______________________ Date __________ Class __________

LESSON 8-8

Reading Strategies

Reading a Table

The formulas below can help you find the surface area of various figures.

Formulas for Finding the Areas of Polygons
Rectangle $A = \ell \cdot w$ ← length • width
Square $A = s^2$ ← side squared, or side • side
Triangle $A = \frac{1}{2}(b \cdot h)$ ← $\frac{1}{2}$ of base • height

The **surface area** of a pyramid is the sum of the areas of all the faces.

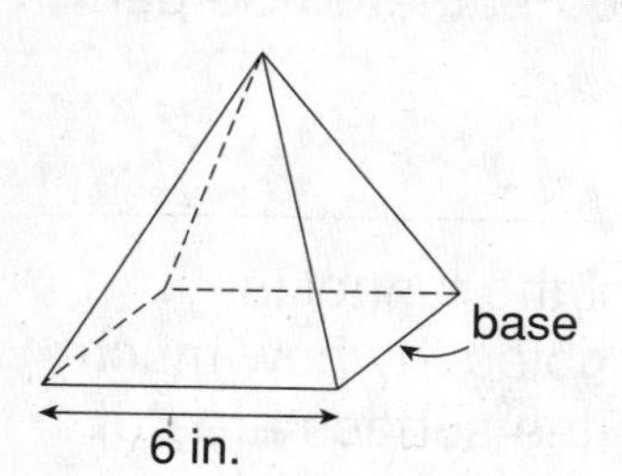

You can use a net to help you choose the formulas to find the surface area of the pyramid.

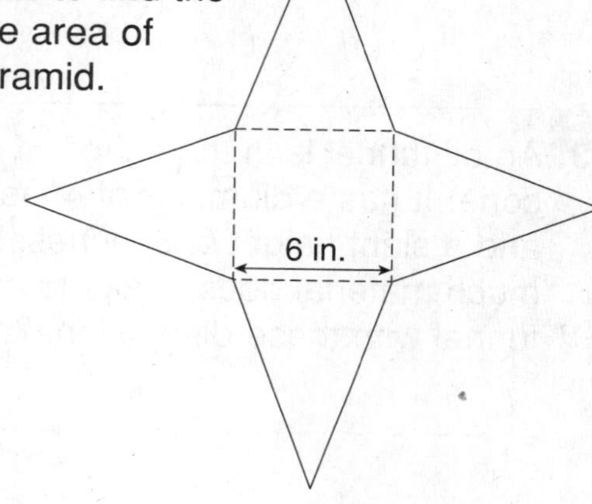

1. Which formula from the table would you choose to find the area of the base?

____

2. One side of the base is 6 inches long. Use this measure to rewrite the formula for the base and compute its area.

____

3. How many faces of the pyramid are made up of triangles? __________

4. Which formula above will you choose to find the area of each triangle? __________

5. The height of each triangle is 8 inches. The base is 6 inches.
   Rewrite the formula with these values. __________

6. Find the area for one triangle. __________

7. Find the surface area for all 4 triangles. __________

8. Combine the area of the base and the four triangles.
   What is the surface area of the pyramid? __________

Name ______________________________ Date ____________ Class ____________

LESSON 8-8

Puzzles, Twisters & Teasers

Dry Up!

Find and circle words from the list in the word search (horizontally, vertically or diagonally). Find a word that answers the riddle. Circle it and write it on the line.

surface	area	slant	height	regular
pyramid	right	cone	lateral	perimeter

```
P Y R A M I D A R E A
E P O L S U R F A C E
R E G U L A R Q W E R
I U O L A T E R A L Y
M I C O N E S L M N K
E T Y U K O I L B H U
T O W E L Y G V A F T
E R D X Z S E W Q N Z
R I G H T H E I G H T
```

What gets wetter as it dries?

A ______________________________

Name ______________________ Date ____________ Class ____________

LESSON 8-9

Practice A

Spheres

Find the volume of each sphere, both in terms of π and to the nearest tenth. $V = \frac{4}{3}\pi r^3$. Use 3.14 for π.

1. $r = 3$ in.

2. $d = 9$ ft

3. $r = 1.5$ m

4. $d = 4$ cm

5. $r = 3.6$ m

6. $d = 10$ cm

Find the surface area of each sphere, both in terms of π and to the nearest tenth. $S = 4\pi r^2$. Use 3.14 for π.

7.

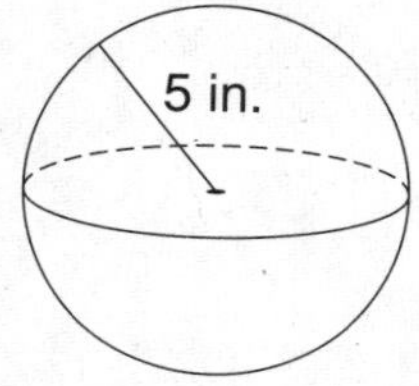

8.

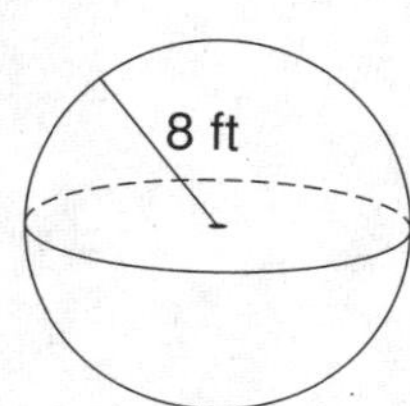

9.

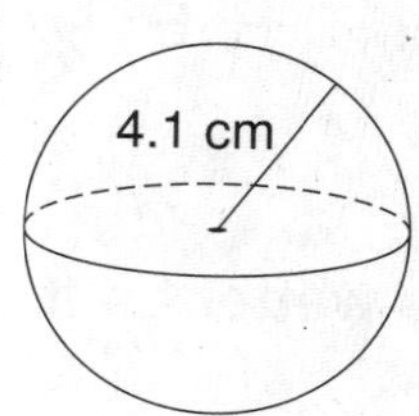

10.

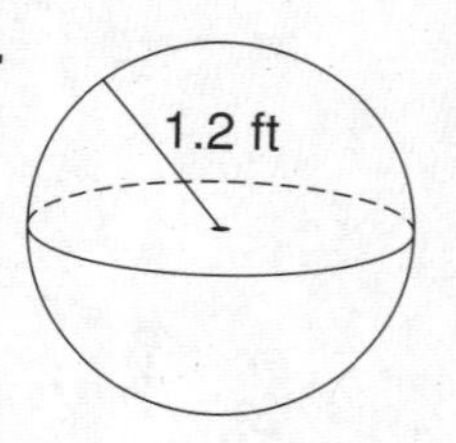

11.

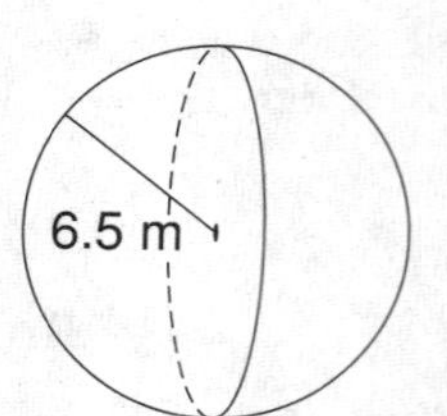

12.

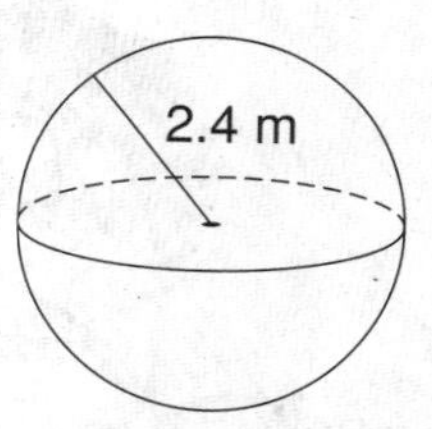

13. A globe is a spherical map of the earth. One of the earliest globes, constructed in 1506 after the discovery of America, is in the New York Public Library. Find the volume and surface area to the nearest tenth of a globe with a diameter of 16 in.

Name ______________________ Date __________ Class __________

LESSON 8-9

Practice B

Spheres

Find the volume of each sphere, both in terms of π and to the nearest tenth. Use 3.14 for π.

1. $r = 9$ ft

2. $r = 21$ m

3. $d = 30$ cm

4. $d = 24$ cm

5. $r = 15.4$ in.

6. $r = 16.01$ ft

Find the surface area of each sphere, both in terms of π and to the nearest tenth. Use 3.14 for π.

7.

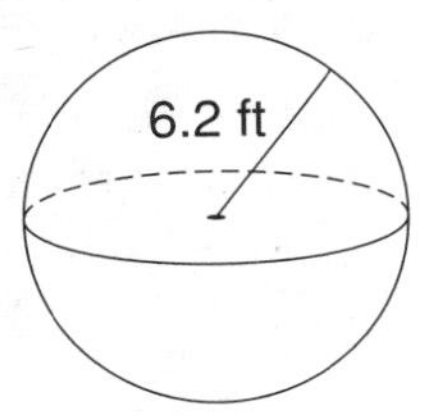

8.

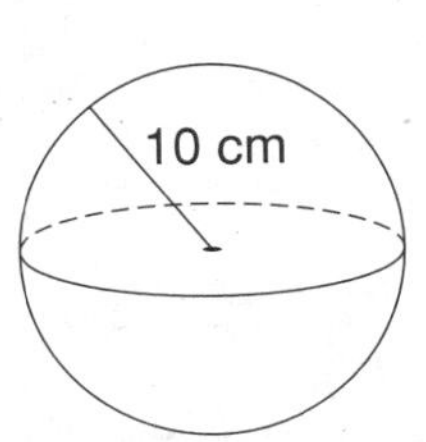

9.

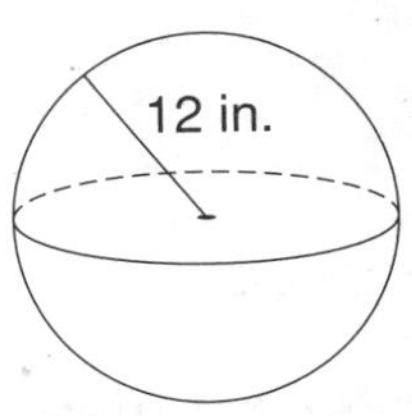

10.

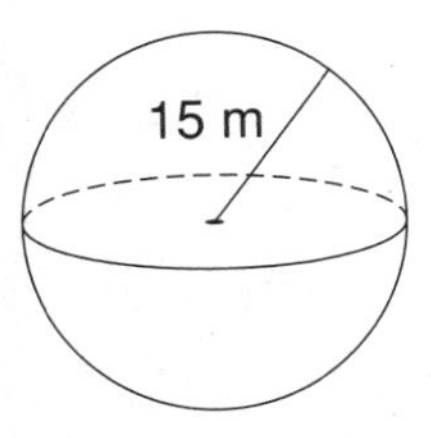

11.

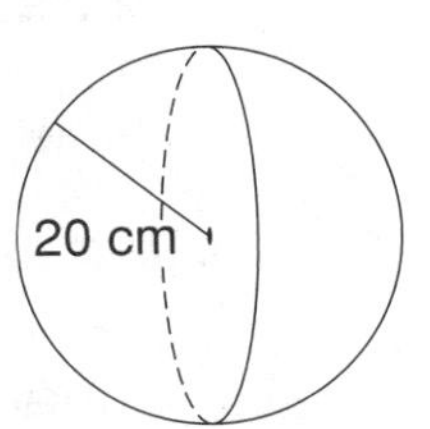

12.

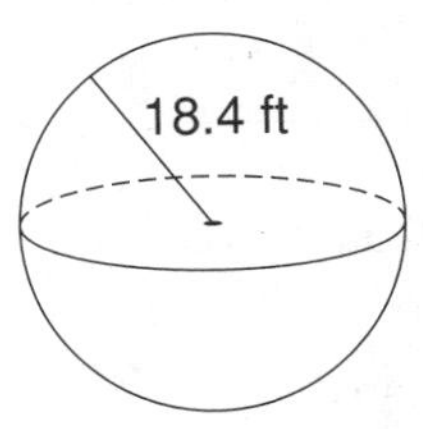

13. In the sport of track and field, a field event is the shot put. This is a game in which a heavy ball or shot is thrown or put for distance. The shot itself comes in various sizes, weights and composition. Find the volume and surface area of a shot with diameter 5.5 cm both in terms of π and to the nearest tenth.

Name ______________________ Date __________ Class __________

LESSON 8-9

Practice C

Spheres

Find the volume of each sphere, both in terms of π and to the nearest tenth. Use 3.14 for π.

1. $r = 6.12$ cm

2. $r = 15$ ft

3. $d = 54$ in.

Find the surface area of each sphere, in terms of π and to the nearest tenth. Use 3.14 for π.

4. 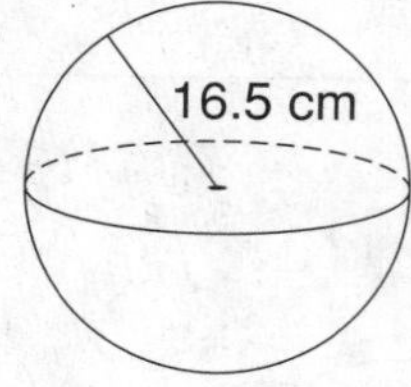

5. 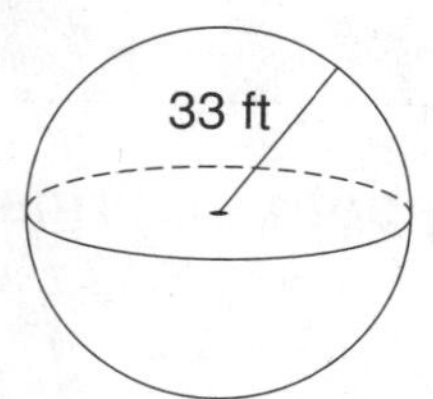

6. 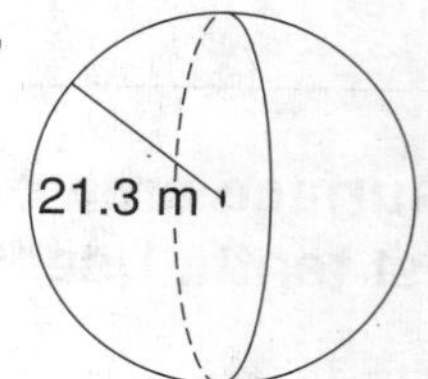

Find the missing measurements of each sphere both in terms of π and to the nearest hundredth. Use 3.14 for π.

7. radius = 8.5 ft
volume = ?
surface area = 289π ft^2

8. radius = 6 m
volume = 288π m^3
surface area = ?

9. diameter = 18 cm
volume = ?
surface area = 324π cm^2

10. radius = ?
diameter = ?
surface area = 576π in^2

11. According to the National Collegiate Athletic Association men's rules, a tennis ball shall have a diameter more than $2\frac{1}{2}$ in. and less than $2\frac{5}{8}$ in. Find the volumes and surface areas of each limit in terms of π.

Name ______________________________ Date ____________ Class ____________

LESSON 8-9

Reteach

Spheres

Sphere: the set of points in space at a fixed distance (its *radius r*) from a fixed point (its *center*)

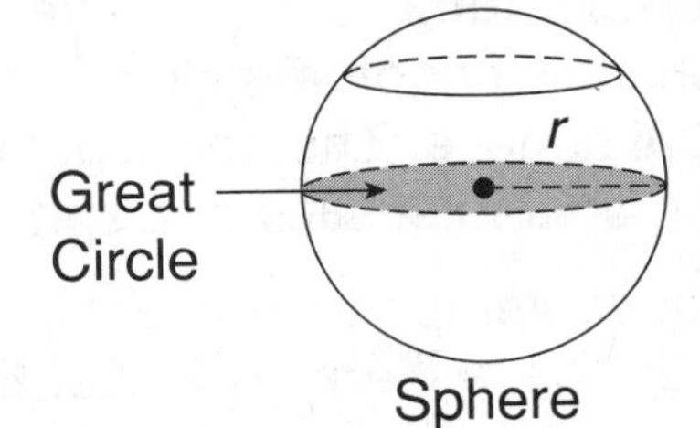

Volume V of Sphere $= \frac{4}{3}\pi r^3$

For a sphere of radius = 9 cm,

$V = \frac{4}{3}\pi r^3$

$V = \frac{4}{3}\pi \times 9^3 = \frac{4}{3}\pi \times 729 = \frac{4 \times \cancel{729}^{243}}{\cancel{3}}\pi = 972\pi \text{ cm}^3$

$V \approx 972(3.14) \approx 3052.08 \text{ cm}^3$

Complete to find the volume of the sphere.

1. $V = \frac{4}{3}\pi r^3$

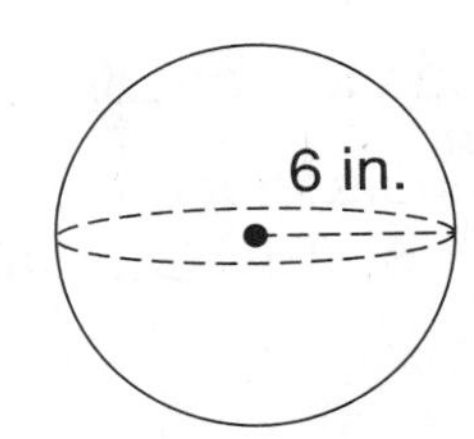

$V = \frac{4}{3}\pi \times$ ______ $= \frac{4}{3}\pi \times$ ______

$V =$ ______ $\pi \text{ in}^3$

$V \approx$ ______ $(3.14) \approx$ __________ in^3

Surface Area S of Sphere $= 4\pi r^2$

For a sphere of radius = 9 in.,

$S = 4\pi r^2$

$S = 4\pi \times 9^2 = 4\pi \times 81 = 324\pi \text{ in}^2$

$S \approx 324(3.14) \approx 1017.36 \text{ in}^2$

Complete to find the surface area of the sphere.

2. $S = 4\pi r^2$

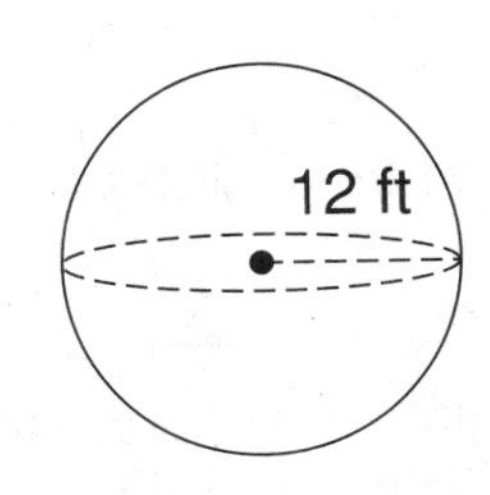

$S = 4\pi \times$ ______ $= 4\pi \times$ ______

$S =$ ______ $\pi \text{ ft}^2$

$S \approx$ ______ $(3.14) \approx$ __________ ft^2

Name ______________________ Date ____________ Class ____________

LESSON 8-9

Challenge

Useful and Intriguing

A **geodesic dome** is a structure made of a complex network of triangles that form a roughly spherical surface. The dome gets its efficiency from the characteristics of a sphere.

The first contemporary geodesic dome (1922) is attributed to the German Walter Bauersfeld. The great-circle principle used in his dome has been used in Asia for centuries to weave fish traps and baskets. In the 1940's, the American Buckminster Fuller used the dome to design efficient houses.

The classic geodesic dome takes its form from the **icosahedron**, a regular solid with 20 equilateral triangles as faces, 30 congruent edges, and 12 vertices.

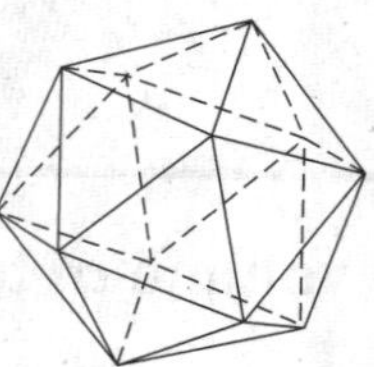

Consider an icosahedron with edge $s = 12$ ft.

1. Find the surface area with the formula Area of Equilateral Triangle $= \frac{s^2}{4}\sqrt{3}$. Use $\sqrt{3} \approx 1.73$ to answer to the nearest tenth of a square foot.

____

2. Find the volume with the formula Volume of Icosahedron $= \frac{5}{12}(3 + \sqrt{5})s^2$. Use $\sqrt{5} \approx 2.24$ to answer to the nearest tenth of a cubic foot.

____

3. Find an approximate value for the radius r of the sphere that has approximately the same volume as the icosahedron.

____

____

4. Using your value of r, find the surface area of that sphere.

____

5. Use your results to make an observation about why a sphere is more efficient than an icosahedron.

____

____

Name ______________________ Date ____________ Class ____________

LESSON 8-9

Problem Solving

Spheres

Early golf balls were smooth spheres. Later it was discovered that golf balls flew better when they were dimpled. On January 1, 1932, the United States Golf Association set standards for the weight and size of a golf ball. The minimum diameter of a regulation golf ball is 1.680 inches. Use 3.14 for π. Round to the nearest hundredth.

1. Find the volume of a smooth golf ball with the minimum diameter allowed by the United States Golf Association.

   ____

2. Find the surface area of a smooth golf ball with the minimum diameter allowed by the United States Golf Association.

   ____

3. Would the dimples on a golf ball increase or decrease the volume of the ball?

   ____

4. Would the dimples on a golf ball increase or decrease the surface area of the ball?

   ____

Use 3.14 for π. Use the following information for Exercises 5–6. A track and field expert recommends changes to the size of a shot put. One recommendation is that a shot put should have a diameter between 90 and 110 mm. Choose the letter for the best answer.

5. Find the surface area of a shot put with a diameter of 90 mm.

   A 25,434 mm^2
   B 101,736 mm^2
   C 381,520 mm^2
   D 3,052,080 mm^2

6. Find the surface area of a shot put with diameter 110 mm.

   F 9,499 mm^2
   G 22,834 mm^2
   H 37,994 mm^2
   J 151,976 mm^2

7. Find the volume of the earth if the average diameter of the earth is 7926 miles.

   A 2.0×10^8 mi^3
   B 2.6×10^{11} mi^3
   C 7.9×10^8 mi^3
   D 2.1×10^{12} mi^3

8. An ice cream cone has a diameter of 4.2 cm and a height of 11.5 cm. One spherical scoop of ice cream is put on the cone that has a diameter of 5.6 cm. If the ice cream were to melt in the cone, how much of it would overflow the cone? Round to the nearest tenth.

   F 0 cm^3 **H** 38.8 cm^3
   G 12.3 cm^3 **J** 54.3 cm^3

Name ______________________ Date __________ Class __________

LESSON 8-9

Reading Strategies

Focus on Vocabulary

A **sphere** is a three-dimensional figure. All the points on the surface of a sphere are the same distance from the center. A basketball is an example of a sphere.

center

radius

A plane that intersects a sphere through its center divides the sphere into two halves called **hemispheres**.

hemisphere

Answer the following questions.

1. How would you describe the location of two points on the surface of a sphere?

____

____

2. What type of a figure is a sphere?

____

3. What divides a sphere into two halves?

____

4. What is each half of a sphere called?

____

5. Give an example of a figure that has the shape of a sphere.

____

Name ______________________ Date __________ Class __________

Puzzles, Twisters & Teasers

Sphere of Influence!

Find the volume of each sphere to the nearest tenth. Use 3.14 for π. Each answer has a corresponding letter. Use the letters to solve the riddle.

1. $V =$ ______________ **L**

2. $V =$ ______________ **H**

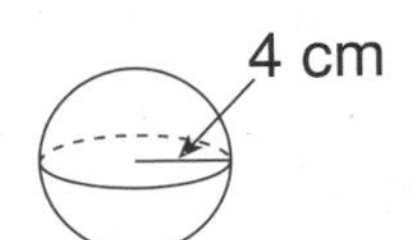

3. $V =$ ______________ **D**

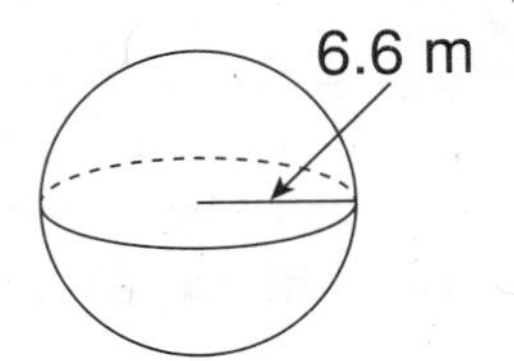

4. $V =$ ______________ **N**

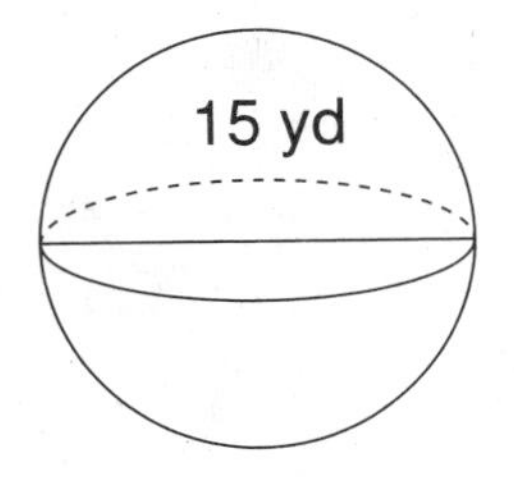

5. $V =$ ______________ **U**

6. $V =$ ______________ **C**

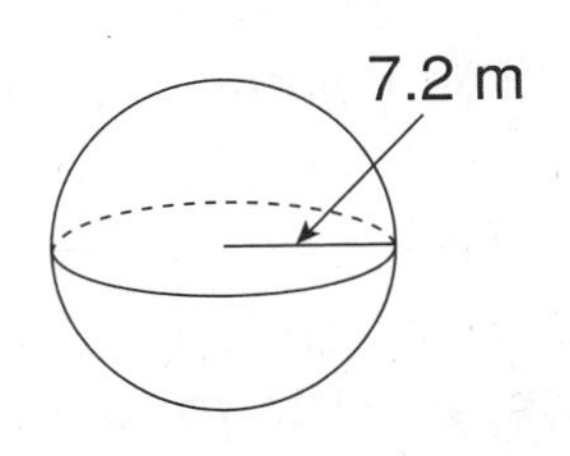

When would you have absolutely no appetite?

Right after _____ _____ _____ _____ _____.

113 65.4 1766.3 1562.7 267.9

Name ______________________ Date ____________ Class ____________

LESSON 8-10

Practice A

Scaling Three-Dimensional Figures

A 3 in. cube is built from small cubes, each 1 in. on a side. Compare the following values.

1. The side lengths of the two cubes

____

____

____

2. The surface area of the two cubes

____

____

3. The volumes of the two cubes

____

A 4 in. cube is built from small cubes, each 2 in. on a side. Compare the following values.

4. The side lengths of the two cubes

____

____

____

5. The surface area of the two cubes

____

____

____

6. The volumes of the two cubes

____

7. The dimensions of an office building are 60 ft long, 80 ft wide, and 160 ft high. The scale model of the office building is 15 in. long. Find the width and height of the model of the office building.

____

Name ______________________ Date __________ Class __________

LESSON **8-10**

Practice B

Scaling Three-Dimensional Figures

A 10 in. cube is built from small cubes, each 2 in. on a side. Compare the following values.

1. The side lengths of the two cubes

____

____

2. The surface area of the two cubes

____

3. The volumes of the two cubes

____

A 9 cm cube is built from small cubes, each 3 cm on a side. Compare the following values.

4. The side lengths of the two cubes

____

____

5. The surface area of the two cubes

____

6. The volumes of the two cubes

____

7. The dimensions of a warehouse are 120 ft long, 180 ft wide, and 60 ft high. The scale model used to build the warehouse is 20 in. long. Find the width and height of the model of the warehouse.

____

8. It takes a machine 40 seconds to fill a cubic box with sides measuring 10 in. How long will it take the same machine to fill a cubic box with sides measuring 15 in.?

____

Name ______________________ Date __________ Class __________

LESSON 8-10

Practice C
Scaling Three-Dimensional Figures

A 16 ft cube is built from small cubes, each 2 ft on a side. Compare the following values.

1. The surface area of the two cubes

____

2. The volumes of the two cubes

____

For each cube, a reduced scale model is built using a scale factor of $\frac{1}{4}$. Find the length of the model and the number of 1 cm cubes used to build it.

3. a 8 cm cube

4. a 16 cm cube

5. a 48 cm cube

6. a 12 cm cube

7. a 20 cm cube

8. a 32 cm cube

9. A 2 ft × 3 ft × 3 ft solid figure is built with 1-ft cubes. If each dimension is doubled, how many more cubes are used to build the larger solid?

____

10. A scale model of an office building is a rectangular prism measuring 10 in. × 15 in. × 26 in. The scale is $\frac{1}{4}$ in. = 1 ft. How many cubic feet of air would be in the empty office building?

____

11. The height of a triangular prism is 16 ft. The sides of the base are 4 ft, $7\frac{1}{2}$ ft, and $8\frac{1}{2}$ ft. The height of a scale model is 2 ft. Find the perimeter of the base of the model.

____

Name ______________________ Date __________ Class __________

LESSON 8-10

Reteach

Scaling Three-Dimensional Figures

Any two cubes are similar.

The sides of this larger cube are 3 times as long as the sides of this smaller cube.

$$\frac{\text{side of larger cube}}{\text{side of smaller cube}} = \frac{9 \text{ in.}}{3 \text{ in.}} = \frac{3}{1} = 3$$

The scale factor is 3.

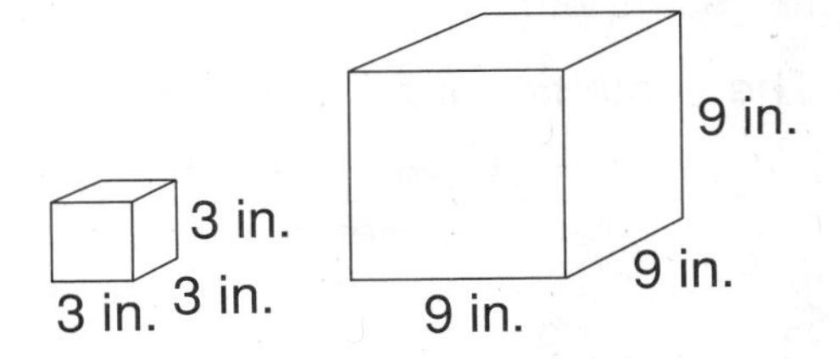

Find the scale factor for each pair of cubes.

1. side of larger cube = 16 cm
   side of smaller cube = 4 cm

   $\frac{\text{larger}}{\text{smaller}} = \frac{16 \text{ cm}}{4 \text{ cm}}$

   = ________

   scale factor = ______

2. side of smaller cube = 9 ft
   side of larger cube = 27 ft

   $\frac{\text{smaller}}{\text{larger}} =$ ________

   = ________

   scale factor = ______

3. side of larger cube = 78 mm
   side of smaller cube = 18 mm

   $\frac{\text{larger}}{\text{smaller}} =$ ________

   = ________

   scale factor = ______

The ratio of the surface areas S of two cubes is the square of the scale factor.

The scale factor for these two cubes is 3.

$$\frac{\text{side of larger cube}}{\text{side of smaller cube}} = \frac{9 \text{ in.}}{3 \text{ in.}} = \frac{3}{1} = 3$$

$$\frac{S \text{ larger}}{S \text{ smaller}} = \frac{6(\text{area one face})}{6(\text{area one face})} = \frac{\overset{1}{\cancel{6}}(\overset{3}{\cancel{9}} \times \overset{3}{\cancel{9}})}{\underset{1}{\cancel{6}}(\underset{1}{\cancel{3}} \times \underset{1}{\cancel{3}})} = \left(\frac{3}{1}\right)^2 = 9$$

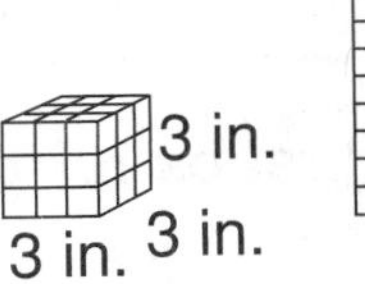

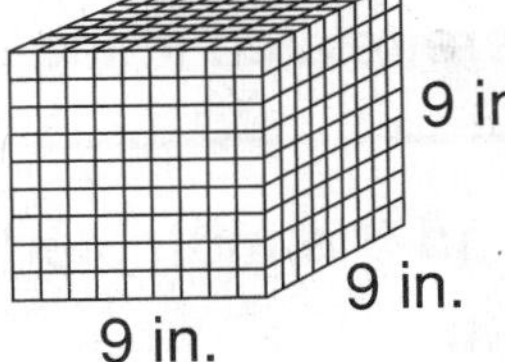

Find the scale factor for each pair of cubes. Then find the ratio of the surface areas.

4. side of smaller cube = 16 in.
   side of larger cube = 64 in.

   $\frac{\text{smaller}}{\text{larger}} =$ ________

   scale factor = ______

   ratio of surface areas

   = (scale factor)^2 = ( )^2 = ________

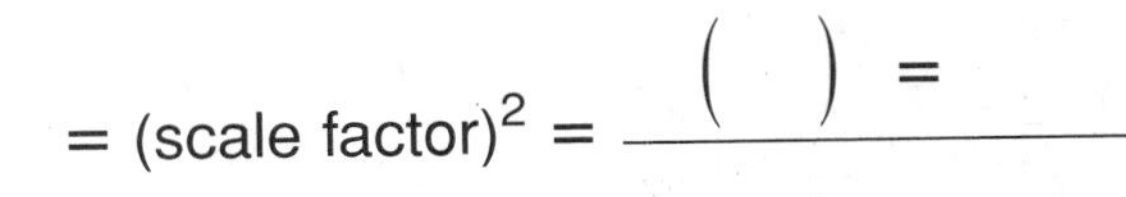

Name ______________________ Date ____________ Class ____________

LESSON 8-10

Reteach

Scaling Three-Dimensional Figures (continued)

The ratio of the volumes V of two cubes is the cube of the scale factor.

The scale factor for these two cubes is 3.

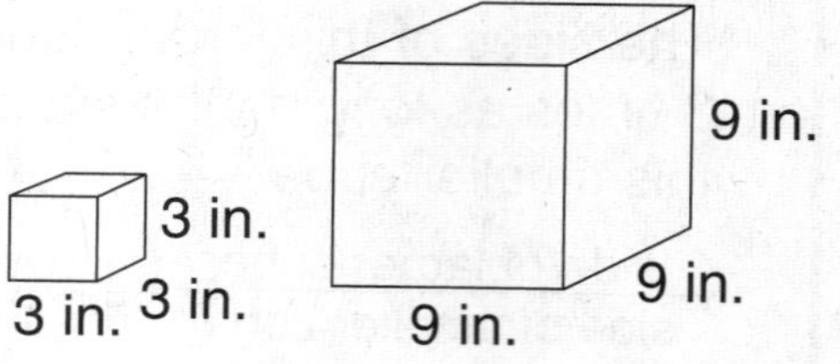

$$\frac{\text{side of larger cube}}{\text{side of smaller cube}} = \frac{9 \text{ in.}}{3 \text{ in.}} = \frac{3}{1} = 3$$

$$\frac{V \text{ larger}}{V \text{ smaller}} = \frac{\ell \times w \times h}{\ell \times w \times h} = \frac{\overset{3}{\cancel{9}} \times \overset{3}{\cancel{9}} \times \overset{3}{\cancel{9}}}{\underset{1}{\cancel{3}} \times \underset{1}{\cancel{3}} \times \underset{1}{\cancel{3}}} = \left(\frac{3}{1}\right)^3 = 27$$

Find the scale factor for each pair of cubes. Then find the ratio of the volumes.

5. side of larger cube = 100 in.
side of smaller cube = 25 in.

$\frac{\text{larger}}{\text{smaller}} = \frac{100 \text{ in.}}{25 \text{ in.}} =$ ______________

scale factor = ______

ratio of volumes

= (scale factor)^3 = (______)^3 = ______

6. side of smaller cube = 6 m
side of larger cube = 36 m

$\frac{\text{smaller}}{\text{larger}} =$ ______________ =

scale factor = ______

ratio of volumes

= (scale factor)^3 = ______ $(\ \ )^3$ =

As with the cube, the measures of other similar solids are related in the same ways to their scale factors.

Find the indicated ratios for these similar cylinders.

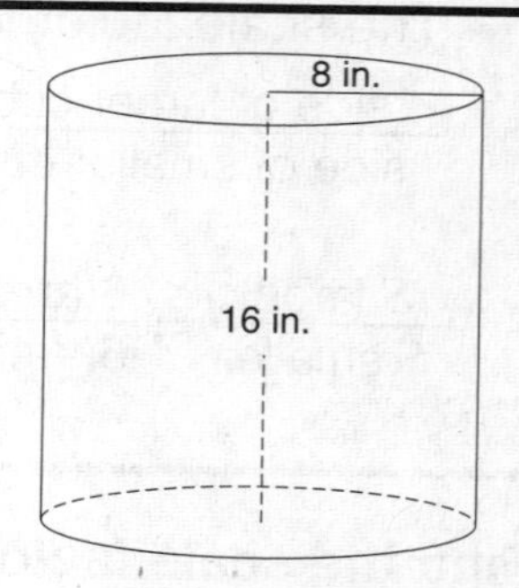

7. $\frac{\text{height of larger cylinder}}{\text{height of smaller cylinder}} =$ ______________ =

8. $\frac{\text{radius of larger cylinder}}{\text{radius of smaller cylinder}} =$ ______________ =

9. scale factor = ______

10. $\frac{\text{area of circular base of larger cylinder}}{\text{area of circular base of smaller cylinder}} = \frac{\pi \bullet (\text{larger radius})^2}{\pi \bullet (\text{smaller radius})^2} =$ ______________ $\frac{\pi \bullet (\ \ )^2}{\pi \bullet (\ \ )^2}$

= ______________ $\frac{\pi}{\pi}$ = ______ = ______ = (scale factor)^2

Name ______________________ Date ____________ Class ____________

LESSON **8-10**

Challenge

Cubed and Diced

A 2 × 2 × 2 cube is painted on all six sides. The cube is cut into eight 1 × 1 × 1 cubes. So, the small cubes are painted on just some of the sides.

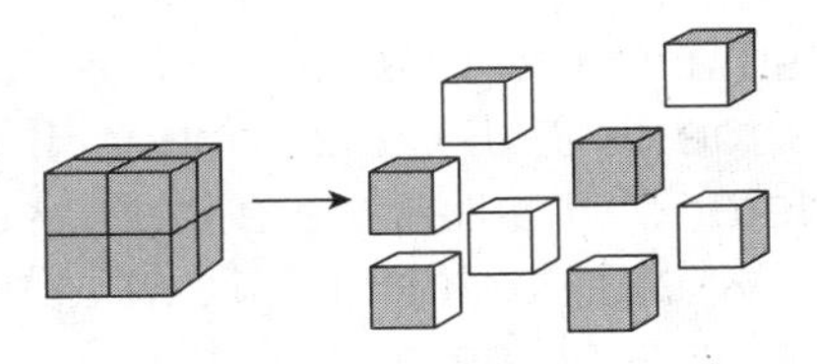

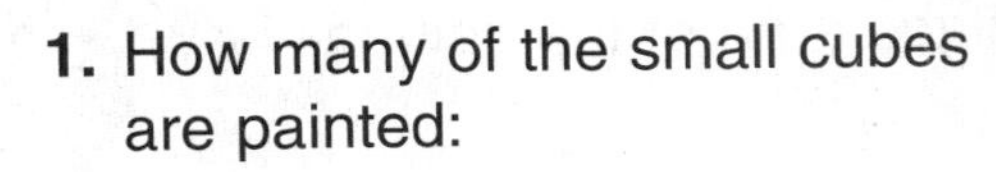

1. How many of the small cubes are painted:

a. on all six sides? ________ **b.** on five sides? ________

c. on four sides? ________ **d.** on three sides? ________

e. on two sides? ________ **f.** on one side? ________

A 3 × 3 × 3 cube is painted on all six sides. The cube is cut into twenty-seven 1 × 1 × 1 cubes.

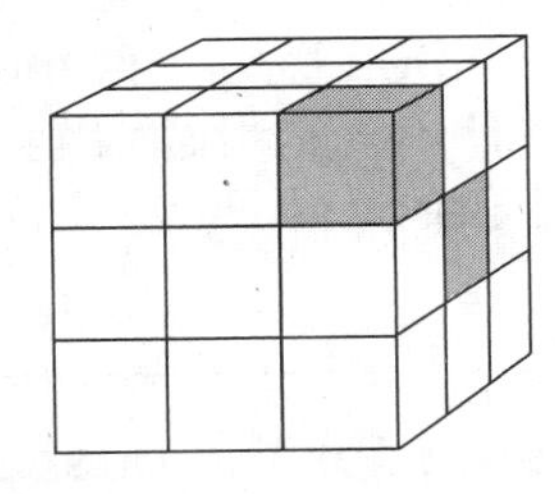

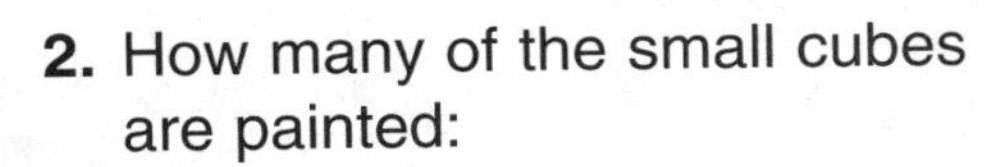

2. How many of the small cubes are painted:

a. on all six sides? ________

b. on five sides? ________ **c.** on four sides? ________

d. on three sides? ________ **e.** on two sides? ________

f. on one side? ________ **g.** on no sides? ________

A 4 × 2 × 1 rectangular prism is painted on all six sides. The prism is cut into eight 1 × 1 × 1 cubes.

3. Draw the figure.

4. How many of the small cubes are painted:

a. on four sides? ________

b. on three sides? ________

c. on two sides? ________

d. on one side? ________

Name ______________________ Date __________ Class __________

LESSON 8-10

Problem Solving

Scaling Three-Dimensional Figures

Round to the nearest hundredth. Write the correct answer.

1. The smallest regulation golf ball has a volume of 2.48 cubic inches. If the diameter of the ball were increased by 10%, or a factor of 1.1, what will the volume of the golf ball be?

____

2. The smallest regulation golf ball has a surface area of 8.86 square inches. If the diameter of the ball were increased by 10%, what will the surface area of the golf ball be?

____

3. The Feathered Serpent Pyramid in Teotihuacan, Mexico, is the third largest in the city. The dimensions of the Sun Pyramid in Teotihuacan, Mexico, are about 3.3 times larger than the Feathered Serpent Pyramid. How many times larger is the volume of the Sun Pyramid than the Feathered Serpent Pyramid?

____

4. The faces of the Feathered Serpent Pyramid and the Sun Pyramid described in Exercise 3 have ancient paintings on them. How many times larger is the surface area of the faces of the Sun Pyramid than the faces of the Feathered Serpent Pyramid?

____

Choose the letter for the best answer.

5. John is designing a shipping container that boxes will be packed into. The container he designed will hold 24 boxes. If he doubles the sides of his container, how many times more boxes will the shipping container hold?

A 2 **C** 8
B 4 **D** 192

6. If John doubles the sides of his container from exercise 5, how many times more material will be required to make the container?

F 2 **H** 8
G 4 **J** 192

7. A child's sandbox is shaped like a rectangular prism and holds 2 cubic feet of sand. The dimensions of the next size sandbox are double the smaller sandbox. How much sand will the larger sandbox hold?

A 4 ft^3 **C** 16 ft^3
B 8 ft^3 **D** 32 ft^3

8. Maria used two boxes of sugar cubes to create a solid building for a class project. She decides that the building is too small and she will rebuild it 3 times larger. How many more boxes of sugar cubes will she need?

F 4 **H** 27
G 25 **J** 52

Name ______________________ Date __________ Class __________

LESSON **8-10**

Reading Strategies

Use Graphic Aids

This cube has a volume of 1 cubic unit. ➝ Volume = 1 $unit^3$
Its surface area is made up of 6 square faces. ➝ Area = 6 $units^2$

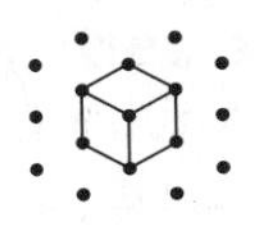

This cube has a volume of 8 cubic units. ➝ Volume = 8 $units^3$
Its surface area is made up of the 4 square units on each of the 6 sides. ➝ $4 \cdot 6 = 24$ $units^2$

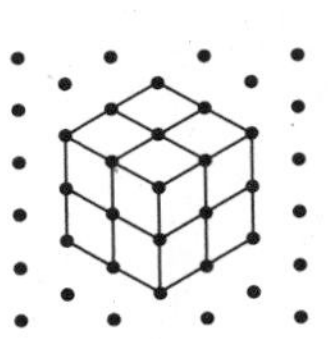

Use this figure to answer the questions.

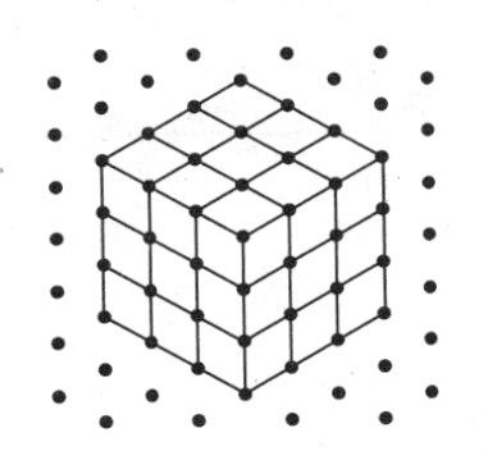

1. What are the dimensions of this cube? ____________

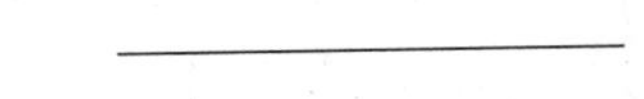

2. How many smaller cubes make up the larger cube? ____________

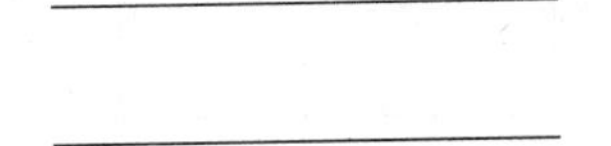

3. What is the volume of this figure? ____________

4. How many square units make up one face of this figure? ____________

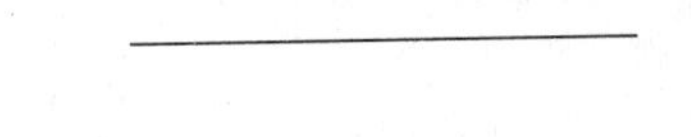

5. How many faces does this figure have? ____________

6. How many square units for all 6 faces of this figure? ____________

Name ______________________ Date ____________ Class ____________

LESSON **8-10**

Puzzles, Twisters & Teasers

Hop To It!

Find the surface area and volume for each cube below. Match your answers with the blanks in the riddle. Fill in the letters to solve the riddle.

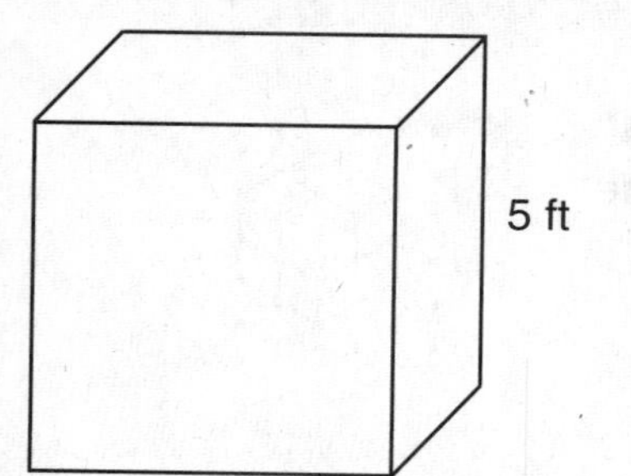

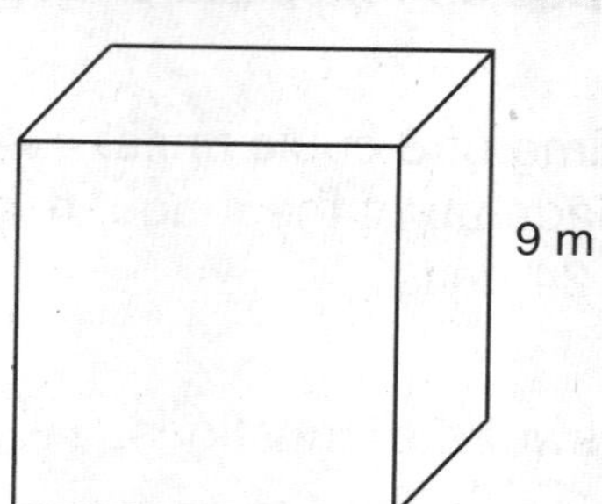

1. surface area = ____________ **R**

2. volume = ____________ **E**

3. surface area = ____________ **I**

4. volume = ____________ **V**

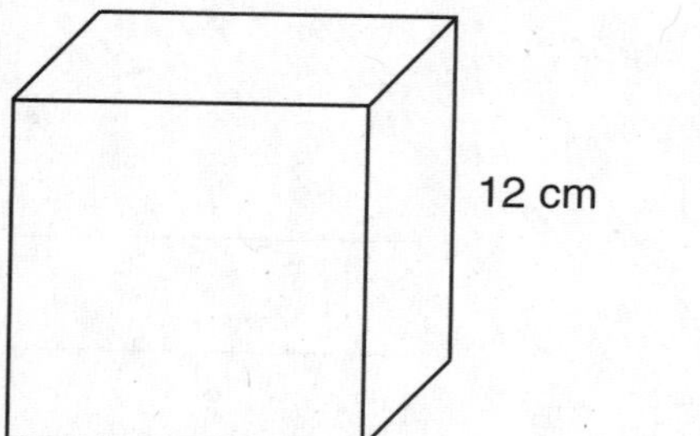

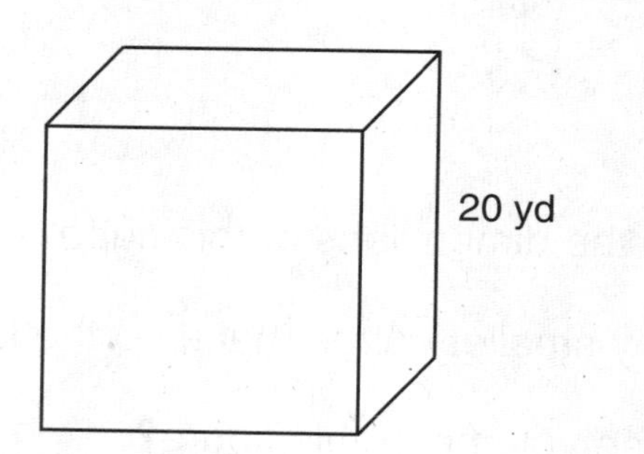

5. surface area = ____________ **G**

6. volume = ____________ **O**

7. surface area = ____________ **H**

8. volume = ____________ **L**

9. surface area = ____________ **S**

10. volume = ____________ **A**

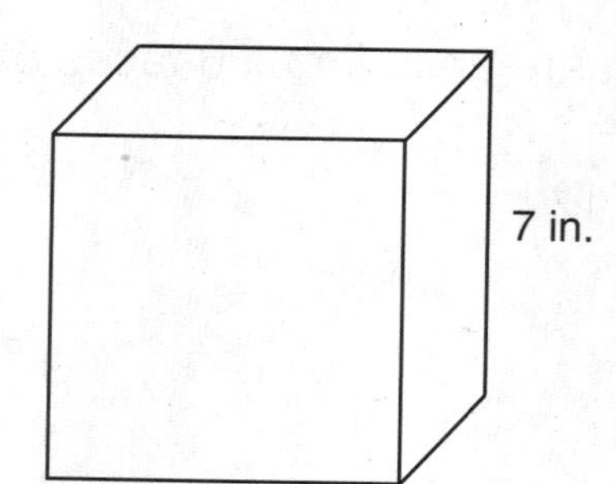

What would you get if you crossed an elephant with a kangaroo?

B ___ ___ ___ ___ ___ ___ ___

486 864 2400 1728 8000 125 294

___ ___ ___ ___ ___ ___ ___ Australia.

343 8000 8000 1728 729 125 150

Name ______________________ Date __________ Class __________

CHAPTER 8

Teacher Tool

Isometric Dot Paper

Name ______________________ Date ____________ Class ____________

CHAPTER 8

Teacher Tool

Cone Net

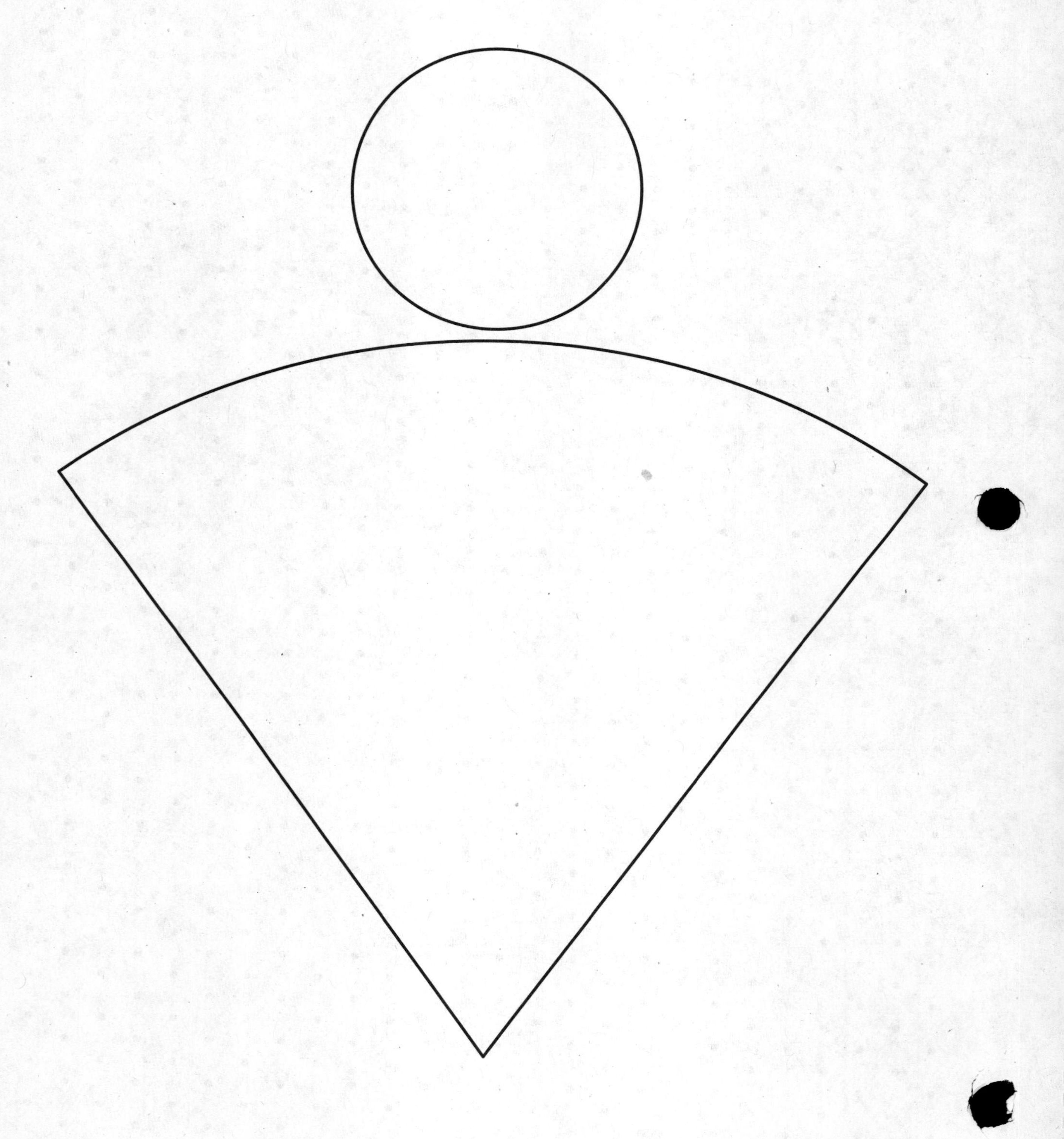

Name ________________________________ Date ______________ Class ______________

CHAPTER 8

Teacher Tool

Pyramid Net

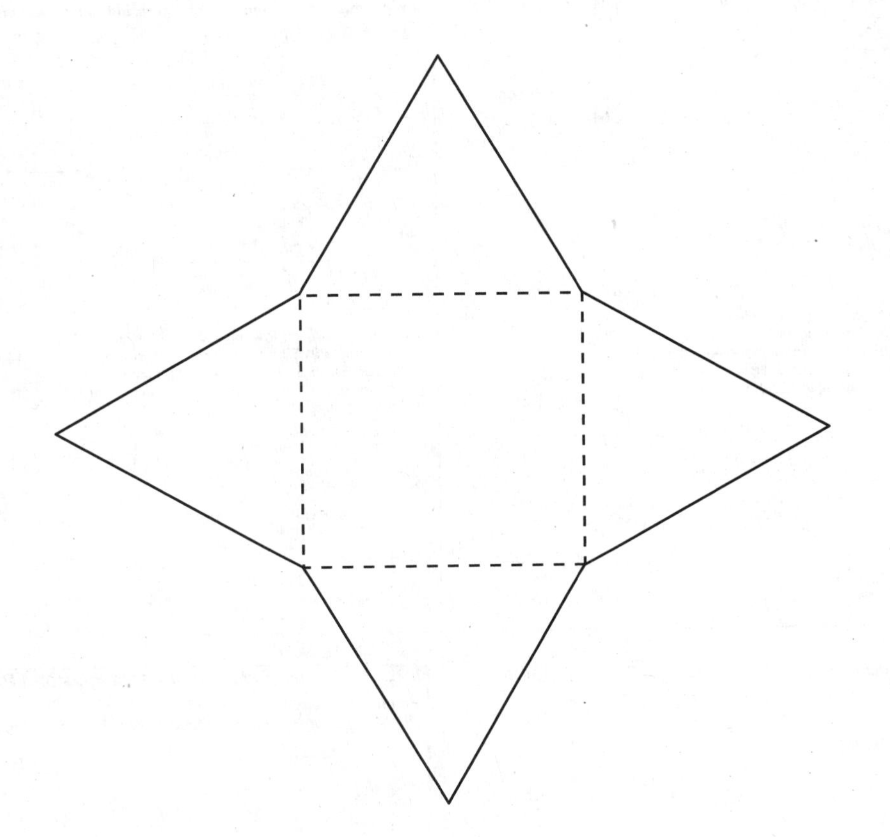

LESSON 8-1

Practice A

Perimeter and Area of Rectangles and Parallelograms

Find the perimeter of each figure.

1. 

18 ft

2. 38 cm

3. 26x in.

Graph and find the area of each figure with the given vertices.

4. (−3, 1), (2, 1), (2, −3), (−3, −3)

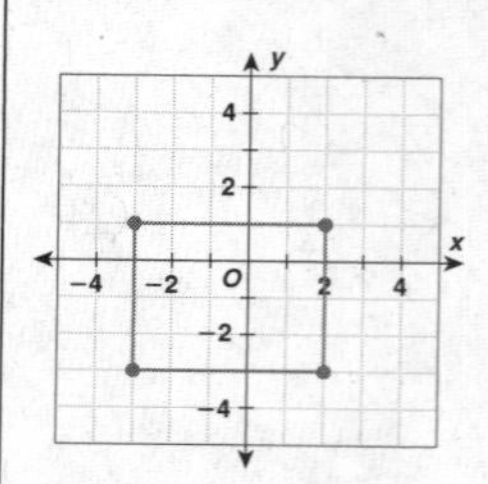

20 units^2

5. (−2, 3), (3, 3), (1, −1), (−4, −1)

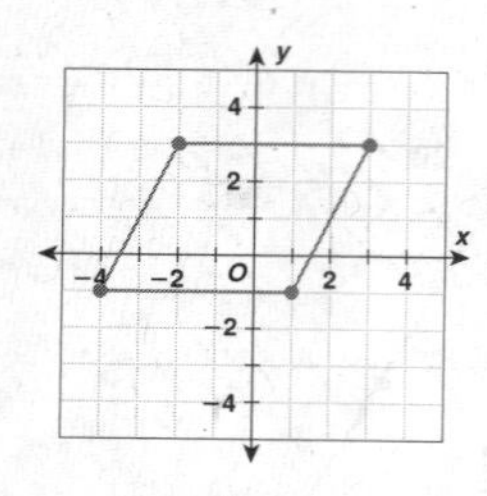

20 units^2

6. The Petersons plan to carpet their new family room. The carpet they have chosen costs $10 per square foot. They also need to install a pad underneath the carpet that costs $2 per square foot. How much will it cost the Petersons to install the carpet and the pad?

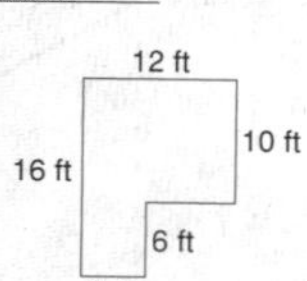

$1800

LESSON 8-1

Practice B

Perimeter and Area of Rectangles and Parallelograms

Find the perimeter of each figure.

1. 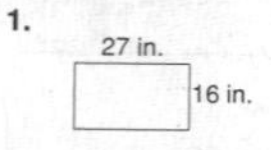

86 in.

2. 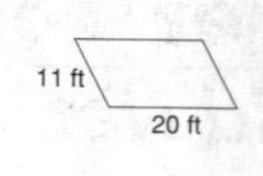

62 ft

3. 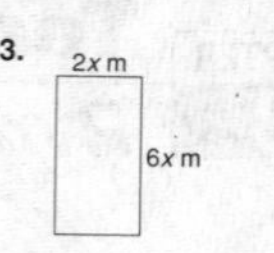

16x m

Graph and find the area of each figure with the given vertices.

4. (−3, 4), (3, 4), (3, −4), (−3, −4)

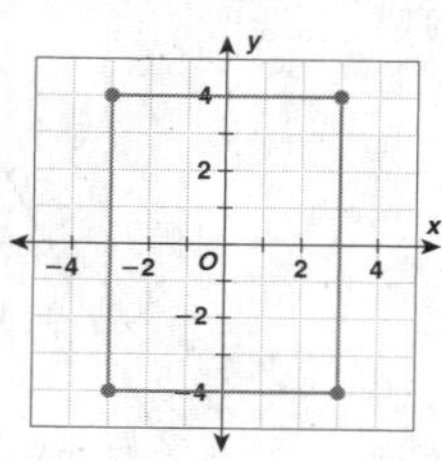

48 units^2

5. (−1, 3), (2, 3), (−1, −4), (−4, −4)

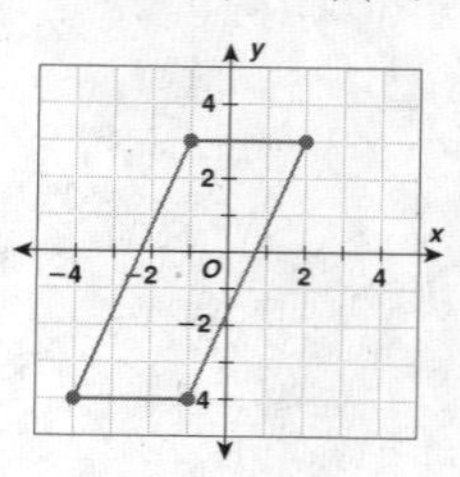

21 units^2

6. Sloppi and Sons Painting Co. charges its customers $1.50 per square foot. How much would Sloppi and Sons charge to paint the rooms of this house if the walls in each room are 9 ft high?

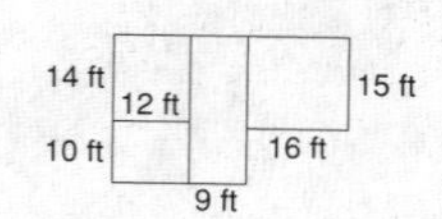

$3024

LESSON 8-1

Practice C

Perimeter and Area of Rectangles and Parallelograms

1. Graph the figure with vertices and find the area of (−1, 3), (4, 3), (3, −4), and (−2, −4).

area = 35 units^2

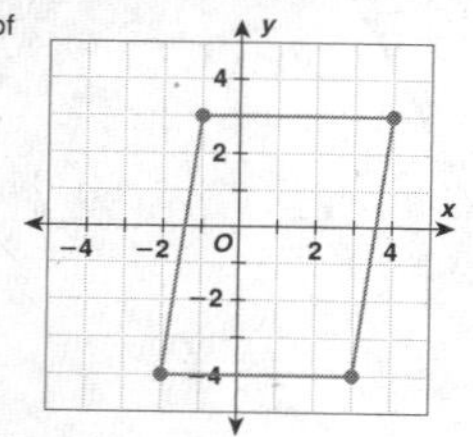

2. Graph the figure with vertices and find the area of (−2, 0), (2, 0), (0, −3), and (−4, −3).

area = 12 units^2

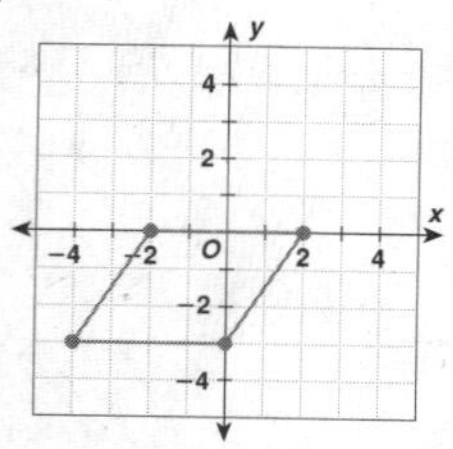

Find the perimeter and area of each figure.

3. 8.5 m, 6 m, 2 m, 4 m, 2 m, 2.5 m, 6.5 m

P = 30 m; A = 25 m^2

4. 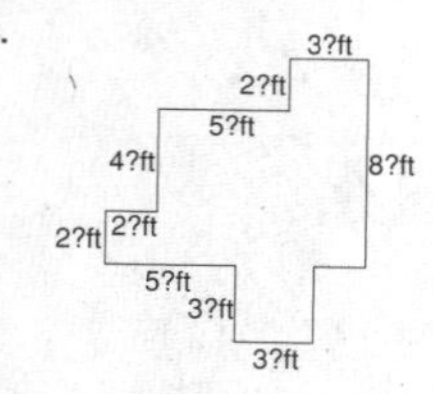

P = 42 ft; A = 67 ft^2

5. The Daughertys bought a piece of land that is 64 ft by 127 ft. They plan on building a two-story house that is 38 ft by 66 ft. How much land will remain after they build the house?

5620 ft^2

LESSON 8-1

Reteach

Perimeter and Area of Rectangles and Parallelograms

Perimeter = distance around a figure.
To find the perimeter of a figure, add the lengths of all its sides.

Perimeter of Rectangle
= b + h + b + h
= **2b + 2h**

Perimeter of Parallelogram
= b + s + b + s
= **2b + 2s**

Complete to find the perimeter of each figure.

1. 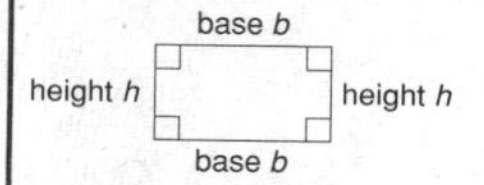

Perimeter of rectangle
= 2b + 2h
= 2(8) + 2(3)
= 16 + 6
= 22 in.

2. 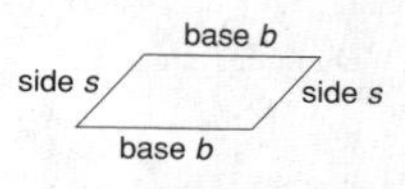

Perimeter of parallelogram
= 2b + 2s
= 2(11) + 2(4)
= 22 + 8
= 30 m

Find the perimeter of each.

3. Large rectangle

P = 8 + 3 + 8 + 3 = 22

4. Small rectangle

P = 4 + 3 + 4 + 3 = 14

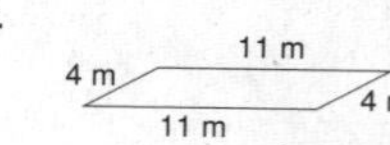

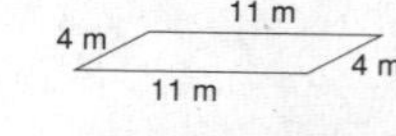

5. The combined rectangles as shown in the figure.

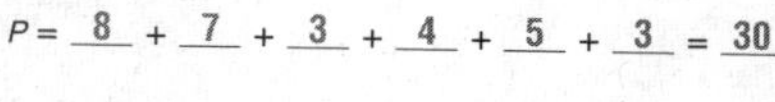

P = 8 + 7 + 3 + 4 + 5 + 3 = 30

LESSON 8-1 **Reteach**

Perimeter and Area of Rectangles and Parallelograms (cont.)

Area = number of square units contained inside a figure.

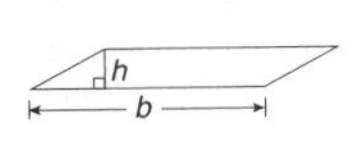

The rectangle contains 12 square units.

Area of rectangle = 4 × 3 = 12 units^2

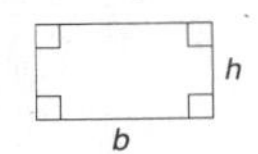

Area of Rectangle = $b \times h$ **Area of Parallelogram** = $b \times h$

Complete to find the area of each figure.

6.

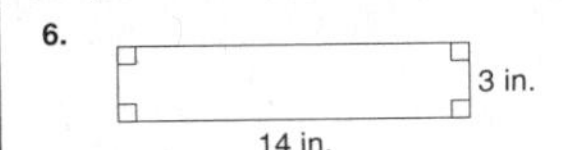

Area of rectangle
= $b \times h$
= 14 × 3 = 42 in^2

7.

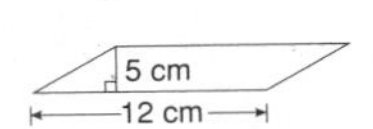

Area of parallelogram *WXYZ*
= $b \times h$
= 12 × 5 = 60 cm^2

8. In the rectangle graphed on the coordinate plane:

base = 2 units

height = 4 units.

Area of rectangle
= base × height
= 2 × 4
= 8 units^2

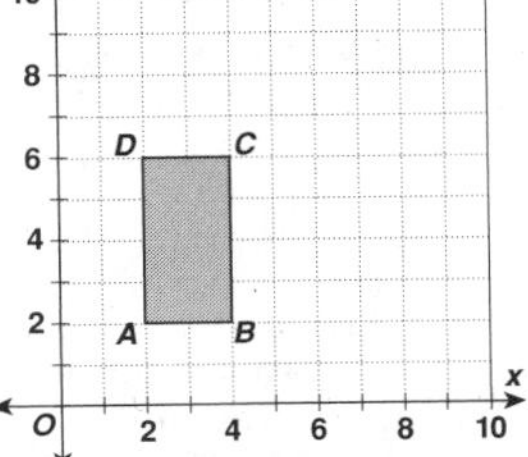

LESSON 8-1 **Challenge**

Color Me Least!

The basic rule for coloring a map is that no two regions that share a boundary can be the same color. However, two regions that meet at only a single point may have the same color.

In 1852, while coloring a map of England, Francis Guthrie noticed that no more than 4 colors were necessary. He conjectured that any map could be colored with no more than four colors.

What came to be known as the **Four Color Map Problem** was considered by mathematicians and school children alike for many years. No satisfactory proof was found until 1976, when K. Appel and W. Haken of the University of Illinois devised a computer program that took 1200 hours to run.

1. a. In this map of 5 distinct regions, can regions *C* and *D* have the same color? Explain.

   No; they share a border.

   b. Can regions *C* and *E* have the same color? Explain.

   Yes; only one point in common.

   c. What is the least number of colors required for this map? 3
   Use numbers to show your answer on the map. Possible answer.

2. Here is a map of 10 neighboring states. So far, as colored, only 3 colors are needed to distinguish among 9 of the 10.
   Color 1: New Mexico, Nevada, and Wyoming
   Color 2: Oregon, Arizona, and Montana
   Color 3: Idaho, Colorado, and California
   How can you color the state of Utah?

   Utah needs a 4th color.

3. Use numbers to color these maps with the least number of colors possible. Possible answer:

LESSON 8-1 **Problem Solving**

Perimeter and Area of Rectangles and Parallelograms

Use the following for Exercises 1–2. A quilt for a twin bed is 68 in. by 90 in.

1. What is the area of the backing applied to the quilt?

   6120 in^2

2. A ruffle is sewn to the edge of the quilt. How many feet of ruffle are needed to go all the way around the edge of the quilt?

   $26\frac{1}{3}$ ft

Use the following for Exercises 3–4. Jaime is building a rectangular dog run that is 12 ft by 8 ft.

3. If the run is cemented, how many square feet will be covered by cement?

   96 ft^2

4. How much fencing will be required to enclose the dog run?

   40 ft

Use the following for Exercises 5–6. Jackie is painting the walls in a room. Two walls are 12 ft by 8 ft, and two walls are 10 ft by 8 ft. Choose the letter for the best answer.

5. What is the area of the walls to be painted?
   (A) 352 ft^2 C 704 ft^2
   B 176 ft^2 D 400 ft^2

6. If a can of paint covers 300 square feet, how many cans of paint should Jackie buy?
   F 1 H 3
   (G) 2 J 4

Use the following for Exercises 7–8. One kind of pool cover is a tarp that stretches over the area of the pool and is tied down on the edge of the pool. The cover extends 6 inches beyond the edge of the pool. Choose the letter for the best answer.

7. A rectangular pool is 20 ft by 10 ft. What is the area of the tarp that will cover the pool?
   A 200 ft^2 C 60 ft^2
   (B) 231 ft^2 D 215.25 ft^2

8. If the tarp costs $2.50 per square foot, how much will the tarp cost?
   F $500.00 H $150.00
   G $538.13 (J) $577.50

LESSON 8-1 **Reading Strategies**

Understanding Symbols in a Formula

Perimeter is the distance around a figure.
The distance around rectangle *A* is
5 inches + 5 inches + 8 inches + 8 inches = 26 inches.

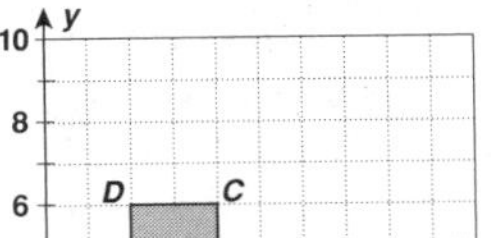

You can find the **p**erimeter of a rectangle by adding 2 times the Base plus 2 times the Height.

Symbols are used to show this. → $P = (2 \cdot b) + (2 \cdot h)$

The symbols make up the **formula** for finding the perimeter of a rectangle.

1. In the formula $P = (2 \cdot b) + (2 \cdot h)$, what does *P* stand for?

   perimeter

2. In the formula $P = (2 \cdot b) + (2 \cdot h)$, what does $(2 \cdot h)$ stand for?

   2 times the height

Area is the amount of surface a figure covers. Area is measured in square units. You can count the square inches in rectangle *C* → 20 square inches.

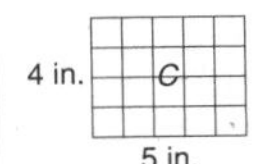

You can find the **a**rea of a rectangle by multiplying the **b**ase times the **h**eight. Symbols are used to show this → $A = b \cdot h$.

The symbols make up the formula for finding the area of a rectangle.

3. In the formula $A = b \cdot h$, what does A stand for? area

4. In the formula $A = b \cdot h$, what does b stand for? length of the base

LESSON 8-1
Puzzles, Twisters & Teasers
What Floats Your Boat?

Find the perimeter of each figure. Each answer has a corresponding letter. Use the letters to solve the riddle.

1. $p =$ **40 m** A
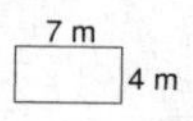

2. $p =$ **20 m** L
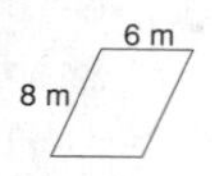

3. $p =$ **36 m** E

4. $p =$ **22 m** I

5. $p =$ **44 m** M
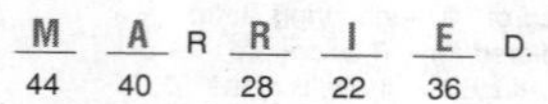

6. $p =$ **28 m** R

You were walking on a bridge and you saw a boat, yet there was not a single person on it. Why?

They were **A** L **L** **M** **A** R **R** **I** **E** D.
(40, 20, 44, 40, 28, 22, 36)

LESSON 8-2
Practice A
Perimeter and Area of Triangles and Trapezoids

Find the perimeter of each figure.

1.
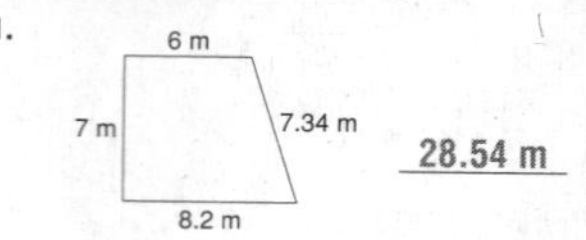

28.54 m

2.
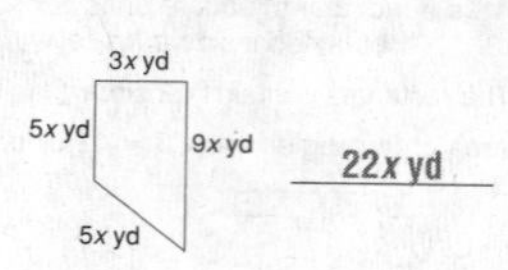

22x yd

Find the missing measurement for each figure with the given perimeter.

3. triangle with perimeter 21 units
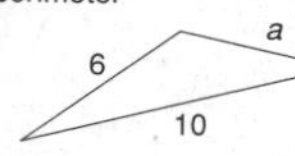

a = 5 units

4. trapezoid with perimeter 39 units
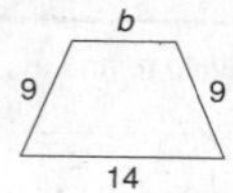

b = 7 units

Graph and find the area of the figure with the given vertices.

5. (−1, 4), (−4, −3), (4, −3)
28 units²
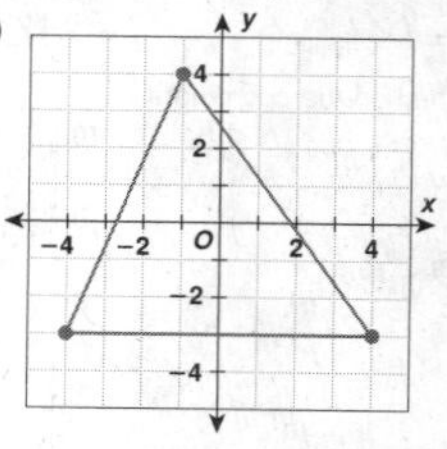
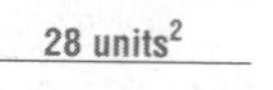

6. A garden shaped like a right triangle has two sides that measure 15 meters and 20 meters. Susie wants to put a fence along the perimeter of the garden.
   a. How long is the third side of the garden? **25 m**
   b. How many meters of fencing material does she need? **60 m**

LESSON 8-2
Practice B
Perimeter and Area of Triangles and Trapezoids

Find the perimeter of each figure.

1.
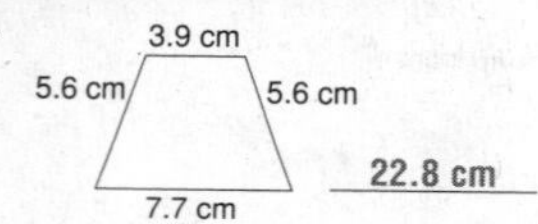

22.8 cm

2.
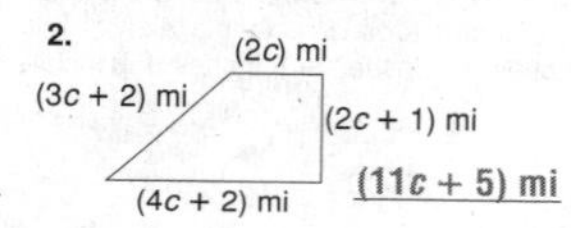

(11c + 5) mi

Find the missing measurement for each figure with the given perimeter.

3. triangle with perimeter 54 units
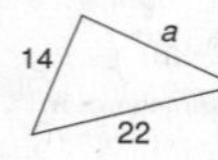

a = 18 units

4. trapezoid with perimeter 34 units
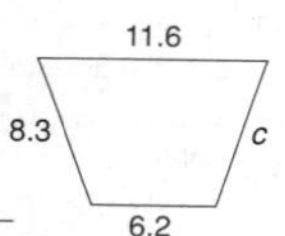

c = 7.9 units

Graph and find the area of each figure with the given vertices.

5. (−1, 3), (4, 3), (4, −4), (−4, −4)

45.5 units²

6. (−1, 2), (−4, −2), (4, −2)
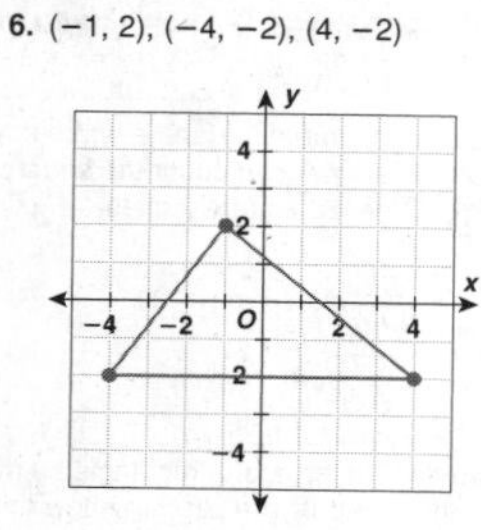
16 units²

7. The two shortest sides of a pennant shaped like a right triangle measure 10 inches and 24 inches. Hank wants to put colored tape around the edge of the pennant. How many inches of tape does he need?
60 in.

LESSON 8-2
Practice C
Perimeter and Area of Triangles and Trapezoids

Graph and find the area of the figure with the given vertices.

1. (−1, 2), (−1, −2), (−3, −4), (−3, 4)
12 units²
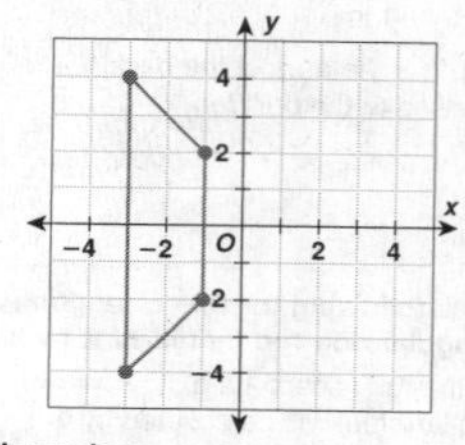

Find the area of each figure with the given dimensions.

2. triangle: $b = 10\frac{2}{3}$, $h = 12\frac{5}{6}$
$68\frac{4}{9}$ units²

3. trapezoid: $b_1 = 15.2$, $b_2 = 7.16$, $h = 14.5$
162.11 units²

4. trapezoid: $b_1 = 9\frac{7}{8}$, $b_2 = 5\frac{3}{4}$, $h = 6\frac{2}{3}$
$52\frac{1}{12}$ units²

5. triangle: $b = 20.5$, $h = 17.64$
180.81 units²

6. The area of a trapezoid is 780 cm². The shorter base is 48 cm and the height is 15 cm. Find the length of the other base.
56 cm

7. Find the height of a triangle with area $85\frac{1}{3}$ ft² and base $10\frac{2}{3}$ ft.
16 ft

8. Find the length of each side of a triangle if the perimeter is 73 in. The length of the second side is twice the length of the first side and the length of the third side is five more than the length of the first side.
first side = 17 in.; second side = 34 in.; third side = 22 in.

9. The two shortest sides of a window shaped like a right triangle measure 24 centimeters and 32 centimeters. Bonnie wants to put a wooden frame around its edges. How many centimeters of wood are needed for the frame?
96 cm

LESSON 8-2
Reteach
Perimeter and Area of Triangles and Trapezoids

To find the perimeter of a figure, add the lengths of all its sides.

Complete to find the perimeter of each figure.

1. 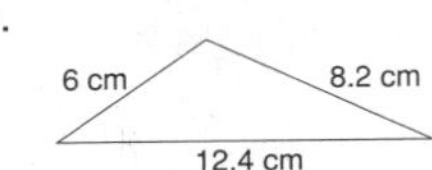

Perimeter of triangle
= 12.4 + 8.2 + 6
= 26.6 cm

2. 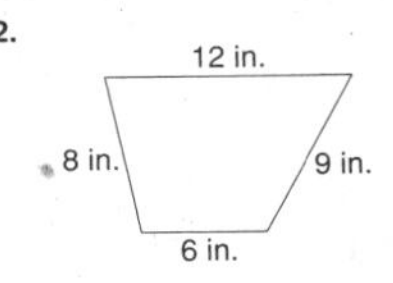

Perimeter of trapezoid
= 6 + 9 + 12 + 8
= 35 in.

Area of Triangle $= \frac{1}{2}bh$
The area of a triangle is one-half the product of a base length b and the height h drawn to that base.

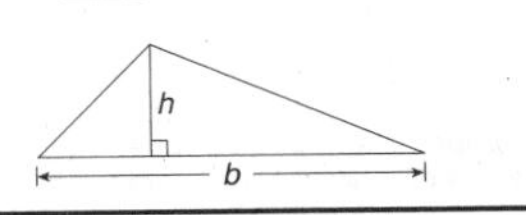

Complete to find the area of each triangle.

3. Area of triangle
$= \frac{1}{2}bh$
$= \frac{1}{2} \times$ 20 $\times$ 4
$= \frac{1}{2} \times$ 80 $=$ 40 in^2

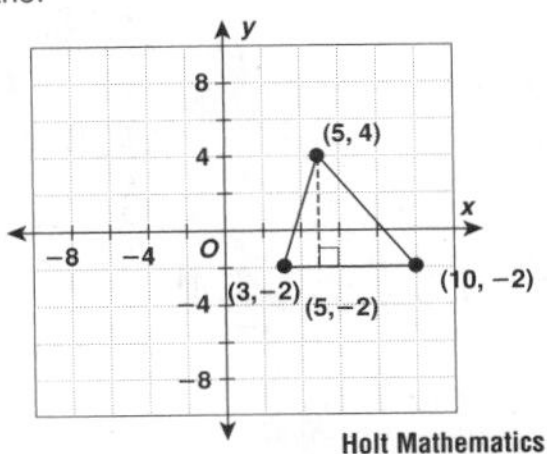

4. In the triangle graphed on the coordinate plane:
base = 10 − 3 = 7 units
height = 4 − (−2) = 6 units.
Area of triangle
$= \frac{1}{2} \times$ base $\times$ height
$= \frac{1}{2} \times$ 7 $\times$ 6
$= \frac{1}{2} \times$ 42 $=$ 21 units^2

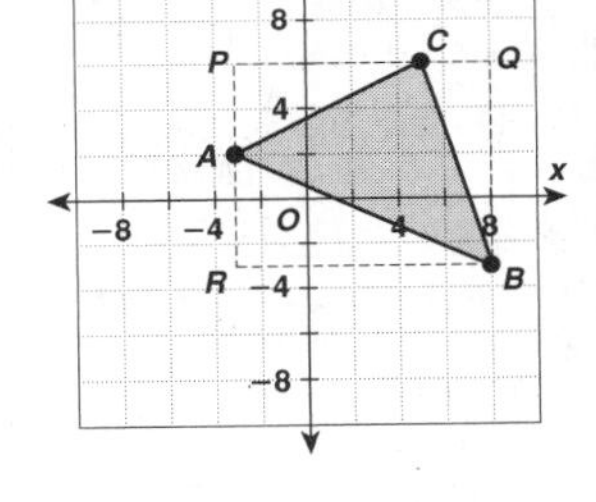

LESSON 8-2
Reteach
Perimeter and Area of Triangles and Trapezoids (continued)

Area of Trapezoid $= \frac{1}{2}h(b_1 + b_2)$
The area of a trapezoid is one-half the height h times the sum of the base lengths b_1 and b_2.

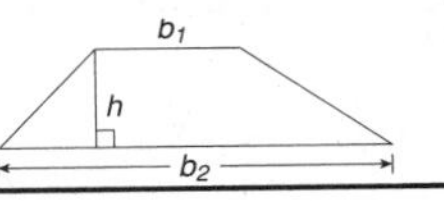

Complete to find the area of each trapezoid.

5. 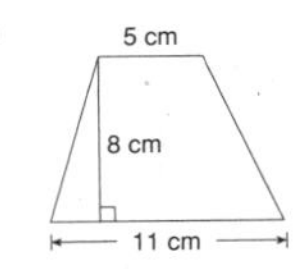

Area of trapezoid
$= \frac{1}{2}h(b_1 + b_2)$
$= \frac{1}{2} \times$ 8 $\times$ (5 + 11)
$=$ 4 $\times$ (16) $=$ 64 cm^2

6. 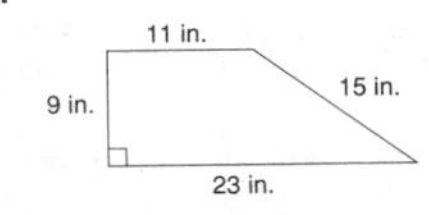

Area of trapezoid
$= \frac{1}{2}h(b_1 + b_2)$
$= \frac{1}{2} \times$ 9 $\times$ (11 + 23)
$=$ 4.5 $\times$ (34) $=$ 153 in^2

7. In the trapezoid graphed on coordinate plane:
$base_1$ = 8 − 4 = 4 units
$base_2$ = 11 − 2 = 9 units
height = 6 − 2 = 4 units.
Area of trapezoid
$= \frac{1}{2} \times$ height $\times$ ($base_1$ + $base_2$)
$= \frac{1}{2} \times$ 4 $\times$ (4 + 9)
$=$ 2 $\times$ (13)
$=$ 26 units^2

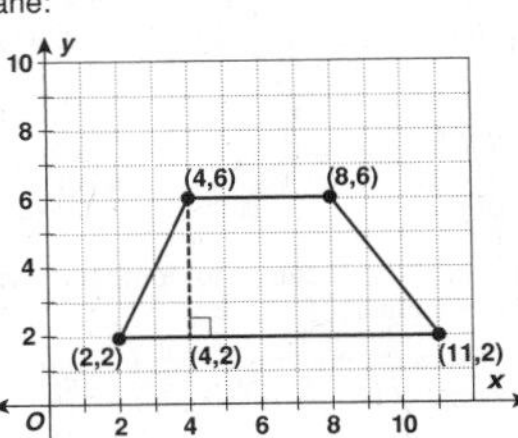

LESSON 8-2
Challenge
Fence Me In!

You can find the area of a triangle in the coordinate plane that has no horizontal or vertical side.

Consider △ABC with vertices $A(-3, 2)$, $B(8, -3)$, and $C(5, 6)$.

By drawing horizontal and vertical lines, △ABC is enclosed in rectangle $PQBR$.

Write the coordinates of the remaining vertices of the rectangle.

1. P (−3, 6)
Q (8, 6)
R (−3, −3)

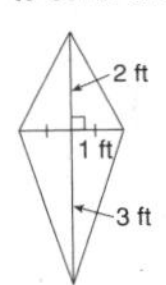

Count boxes or subtract coordinates to find the indicated dimensions. Then find the indicated areas.

2. For rectangle $PQBR$:
base RB = 11 units height RP = 9 units area = 99 units^2

3. For right triangle APC:
base PC = 8 units height AP = 4 units area = 16 units^2

4. For right triangle CQB:
base CQ = 3 units height BQ = 9 units area = 13.5 units^2

5. For right triangle ARB:
base RB = 11 units height AR = 5 units area = 27.5 units^2

6. Explain how to combine the areas of the rectangle and the three right triangles to find the area of △ABC. Then find the area of △ABC.

△ABC = rectangle $PQBR$ − (△APC + △CQB + △ARB)
△ABC = 99 − (16 + 13.5 + 27.5) = 42 units^2

LESSON 8-2
Problem Solving
Perimeter and Area of Triangles and Trapezoids

Write the correct answer.

1. Find the area of the material required to cover the kite pictured below.

5 ft^2

2. Find the area of the material required to cover the kite pictured below.

9 ft^2

3. Find the approximate area of the state of Nevada.

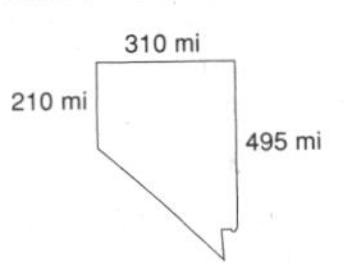

109,275 mi^2

4. Find the area of the hexagonal gazebo floor.

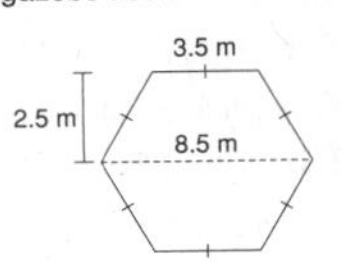

30 m^2

Choose the letter for the best answer.

5. Find the amount of flooring needed to cover the stage pictured below.

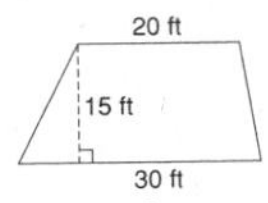

A 4500 ft^2
B 750 ft^2
C 525 ft^2
Ⓓ 375 ft^2

6. Find the combined area of the congruent triangular gables.

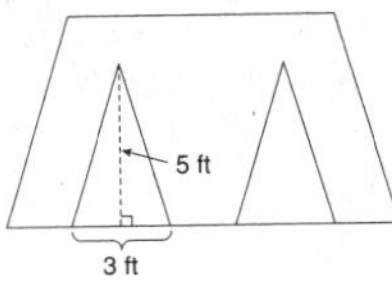

F 7.5 ft^2
Ⓖ 15 ft^2
J 60 ft^2
H 30 ft^2

LESSON **8-2**

Reading Strategies

Compare and Contrast

Compare methods for finding the perimeter of rectangles and triangles.

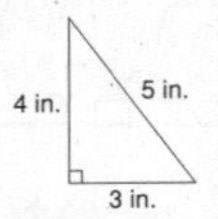

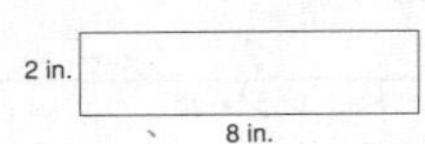

Perimeter of Triangles
Add lengths of all 3 sides.
$P = a + b + c$

Perimeter of Rectangles
Add lengths of all 4 sides.
$P = b + h + b + h$

Answer the questions to compare finding the perimeter of rectangles and triangles.

1. How is finding the perimeter of triangles and rectangles alike?

   You add the lengths of all the sides.

2. How is finding the perimeters of these figures different?

   There are 4 sides to add for rectangles and only 3 sides to add for triangles.

Compare methods for finding the area of rectangles and triangles.

Area of Rectangles
$A = \text{base} \cdot \text{height}$
$A = b \cdot h$

Area of Triangles
$A = \frac{1}{2} \cdot \text{base} \cdot \text{height}$
$A = \frac{1}{2}bh$

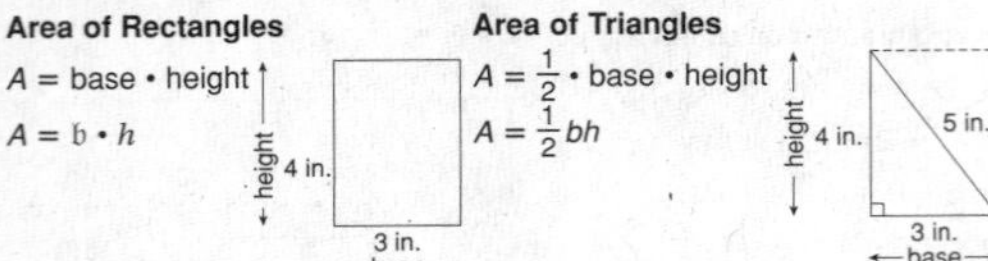

Answer the questions to compare finding the area of rectangles and triangles.

3. How does the area of this triangle compare to the area of this rectangle?

   The triangle has half the area of the rectangle.

4. How is finding the area of this triangle different than finding the area of this rectangle?

   You use base and height instead of length and width. The area of the triangle is only one-half the base times the height.

LESSON **8-2**

Puzzles, Twisters & Teasers

A Whole Hole!

Find the perimeter of each figure. Each answer has a corresponding letter. Use the letters to solve the riddle.

1. $p =$ 32 ft **E**
2. $p =$ 24 ft **M**
3. $p =$ 12 ft **A**

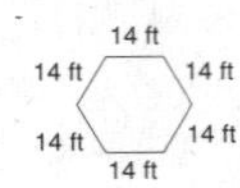

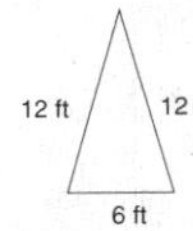

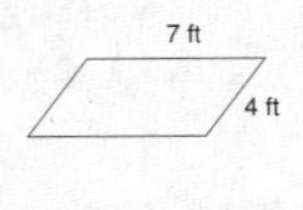

4. $p =$ 84 ft **H**
5. $p =$ 30 ft **K**
6. $p =$ 22 ft **O**

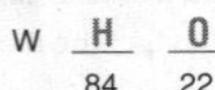

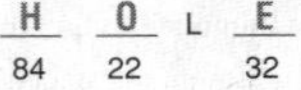

14 ft, 14 ft, 14 ft, 14 ft, 14 ft, 14 ft (hexagon); 12 ft, 12 ft, 6 ft (triangle); 7 ft, 4 ft (parallelogram)

Suppose you were in a room with no door or windows, but two halves of a table. How could you get out?

Put the halves together to M A K E A W H O L E

M	A	K	E	A	W	H	O	L	E
24	12	30				84	22		32

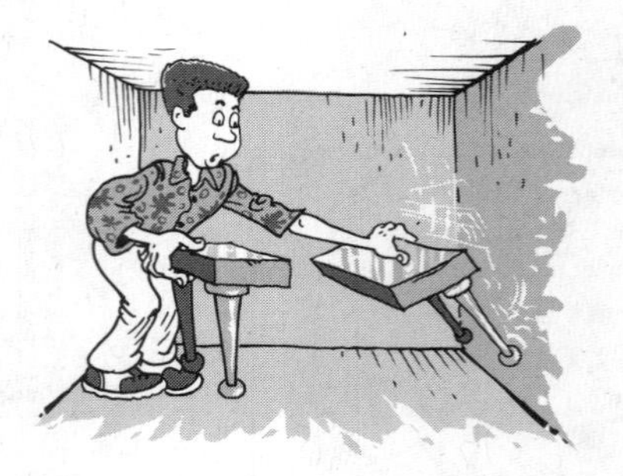

LESSON **8-3**

Practice A

Circles

Find the circumference of each circle, both in terms of π and to the nearest tenth. $C = \pi d$. Use 3.14 for π.

1. circle with diameter 4 cm

   4π cm or 12.6 cm

2. circle with radius 4 in.

   8π in. or 25.1 in.

3. circle with radius 5.5 ft

   11π ft or 34.5 ft

4. circle with diameter 3 m

   3π m or 9.4 m

5. circle with radius 9 cm

   18π cm or 56.5 cm

6. circle with diameter 8.2 in.

   8.2π in. or 25.7 in.

Find the area of each circle, both in terms of π and to the nearest tenth. $A = \pi r^2$. Use 3.14 for π.

7. circle with radius 5 m

   $25\pi\ m^2$ or $78.5\ m^2$

8. circle with diameter 8 ft

   $16\pi\ ft^2$ or $50.2\ ft^2$

9. circle with diameter 6 cm

   $9\pi\ cm^2$ or $28.3\ cm^2$

10. circle with radius 9 in.

    $81\pi\ in^2$ or $254.3\ in^2$

11. circle with radius 3.1 m

    $9.61\pi\ m^2$ or $30.2\ m^2$

12. circle with diameter 4.8 cm

    $5.76\pi\ cm^2$ or $18.1\ cm^2$

13. Graph a circle with center (0, 0) that passes through (0, 2). Find the area and circumference, both in terms of π and to the nearest tenth. Use 3.14 for π.

    $A = 4\pi\ units^2$ or $12.6\ units^2$;
    $C = 4\pi$ units or 12.6 units

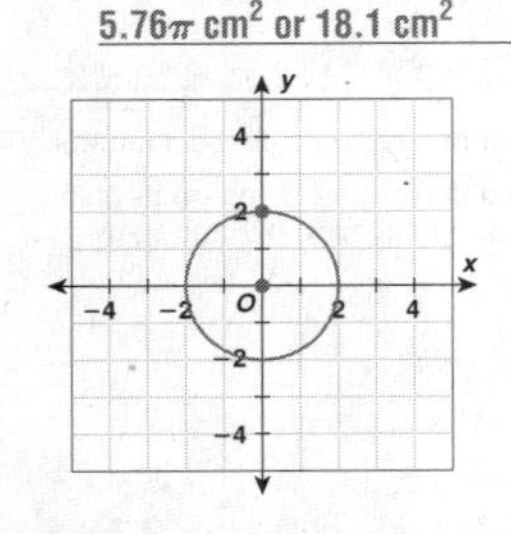

14. A wheel has a diameter of 21 inches. About how far does it travel if it makes 30 complete revolutions? Use $\frac{22}{7}$ for π.

    1,980 in.

LESSON **8-3**

Practice B

Circles

Find the circumference of each circle, both in terms of π and to the nearest tenth. Use 3.14 for π.

1. circle with radius 10 in.

   20π in. or 62.8 in.

2. circle with diameter 13 cm

   13π cm or 40.8 cm

3. circle with diameter 18 m

   18π m or 56.5 m

4. circle with radius 15 ft

   30π ft or 94.2 ft

5. circle with radius 11.5 in.

   23π in. or 72.2 in.

6. circle with diameter 16.4 cm

   16.4π cm or 51.5 cm

Find the area of each circle, both in terms of π and to the nearest tenth. Use 3.14 for π.

7. circle with radius 9 in.

   $81\pi\ in^2$ or $254.3\ in^2$

8. circle with diameter 14 cm

   $49\pi\ cm^2$ or $153.9\ cm^2$

9. circle with radius 20 ft

   $400\pi\ ft^2$ or $1256\ ft^2$

10. circle with diameter 17 m

    $72.3\pi\ m^2$ or $226.9\ m^2$

11. circle with diameter 15.4 m

    $59.3\pi\ m^2$ or $186.2\ m^2$

12. circle with radius 22 yd

    $484\pi\ yd^2$ or $1519.8\ yd^2$

13. Graph a circle with center (0, 0) that passes through (0, −3). Find the area and circumference, both in terms of π and to the nearest tenth. Use 3.14 for π.

    $A = 9\pi\ units^2$ or $28.3\ units^2$;
    $C = 6\pi$ units or 18.8 units

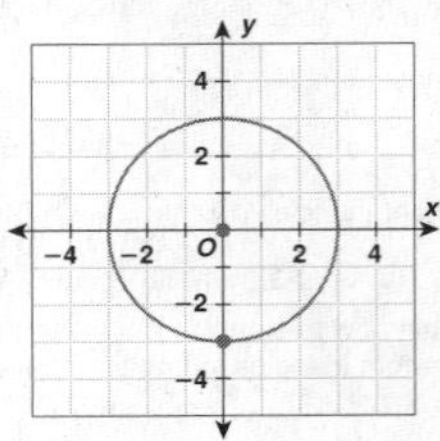

14. A wheel has a radius of $2\frac{1}{3}$ feet. About how far does it travel if it makes 60 complete revolutions? Use $\frac{22}{7}$ for π.

    880 ft

LESSON 8-3
Practice C
Circles

Find the circumference and area of each circle to the nearest tenth. Use 3.14 for π.

1. 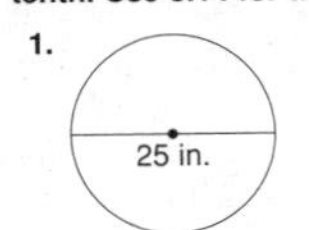

$C = 78.5$ in.

$A = 490.6$ in^2

2. 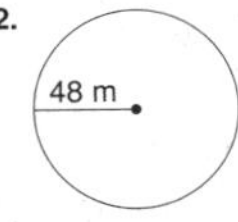

$C = 301.4$ m

$A = 7234.6$ m^2

3. 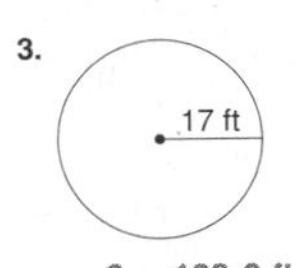

$C = 106.8$ ft

$A = 907.5$ ft^2

Find the radius of each circle with the given measurement.

4. $C = 36\pi$ cm — 18 cm
5. $C = 42\pi$ ft — 21 ft
6. $C = 80.1\pi$ m — 40.05 m
7. $C = 152.6\pi$ in. — 76.3 in.
8. $A = 121\pi$ cm^2 — 11 cm
9. $A = 734.41\pi$ ft^2 — 27.1 ft
10. $A = 1024\pi$ m^2 — 32 m
11. $A = 5184\pi$ in^2 — 72 in.

Find the shaded area to the nearest tenth. Use 3.14 for π.

12. 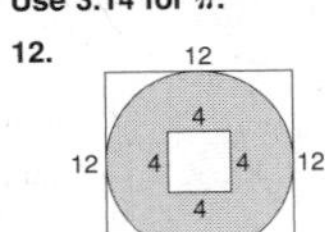

97.0 units^2

13. 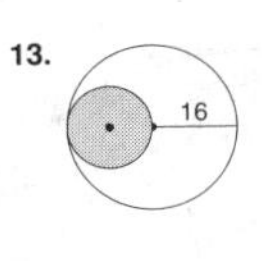

201.0 units^2

14. 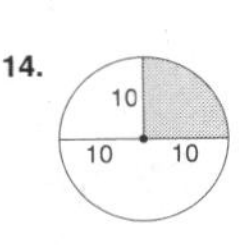

78.5 units^2

15. Derek is riding his bicycle with a 36-in. diameter wheel in a 10-mi race. How many revolutions will each wheel make in the race? Round your answer to the nearest tenth. Use 3.14 for π. (Hint: 12 in. = 1 ft; 5280 ft = 1 mi)

5605.1 revolutions

CHAPTER 8-3
Reteach
Circles

A **radius** connects the **center** of a **circle** to any point on the circle.

A **diameter** passes through the center and connects two points on the circle.

diameter d = twice radius r
$d = 2r$

Circumference — Center — Radius — Radius — Diameter

Circumference is the distance around a circle.

(The symbol $\approx$ means *is approximately equal to.*)

Circumference $C \approx 3$(diameter d)
$C = \pi d$

For a circle with diameter = 8 in.
$C = \pi d$
$C = \pi(8)$
$C = 8\pi$ in.
$\pi \approx 3.14$ $C \approx 8(3.14) \approx 25.12$ in.

Circumference $C \approx 6$(radius r)
$C = 2\pi r$

For a circle with radius = 8 in.
$C = 2\pi r$
$C = 2\pi(8)$
$C = 16\pi$ in.
$\pi \approx 3.14$ $C \approx 16(3.14) \approx 50.24$ in.

Find the circumference of each circle, exactly in terms of π and approximately when $\pi = 3.14$.

1. diameter = 15 ft
$C = \pi d$
$C = \pi($ 15 $) =$ 15π ft
$C \approx 3.14($ 15 $) \approx$ 47.1 ft

2. radius = 4 m
$C = 2\pi r$
$C = 2\pi($ 4 $) =$ 8π m
$C \approx$ 8 $(3.14) \approx$ 25.12 m

Area $A \approx 3$(the square of radius r)
$A = \pi r^2$
For a circle with radius = 5 in.: $A = \pi r^2 = \pi(5^2) = 25\pi$ in^2
$A \approx 25(3.14) \approx 78.5$ in^2

Find the area of each circle, exactly in terms of π and approximately when $\pi = 3.14$.

3. radius = 9 ft
$A = \pi r^2$
$A = \pi($ 9^2 $) =$ 81π ft^2
$A \approx$ 81 $(3.14) \approx$ 254.34 ft^2

4. diameter = 10 m, radius = 5 m
$A = \pi r^2$
$A = \pi($ 5^2 $) =$ 25π m^2
$A \approx$ 25 $(3.14) \approx$ 78.5 m^2

LESSON 8-3
Challenge
Circles

Work in $\triangle PQR$ at the bottom of this page.

1. Use a ruler to find the midpoints of the three sides of the triangle. Label these midpoints M_1, M_2, M_3.
2. An **altitude** of a triangle is a segment drawn from a vertex perpendicular to the opposite side. Draw the altitude to each side of the triangle. Use F_1, F_2, F_3 to label the foot of each altitude (where the altitude meets the side of the triangle at right angles).
3. The point at which the altitudes meet is the **orthocenter** of the triangle. Label the orthocenter O. Locate the midpoints of $\overline{OP}$, $\overline{OQ}$, $\overline{OR}$, the segments that connect orthocenter O to each vertex. Labels these midpoints D_1, D_2, D_3.
4. If you have been accurate in your measurements, the nine points – M_1, M_2, M_3, F_1, F_2, F_3, D_1, D_2, D_3 – lie on a circle. Draw the **nine-point circle** for $\triangle PQR$.

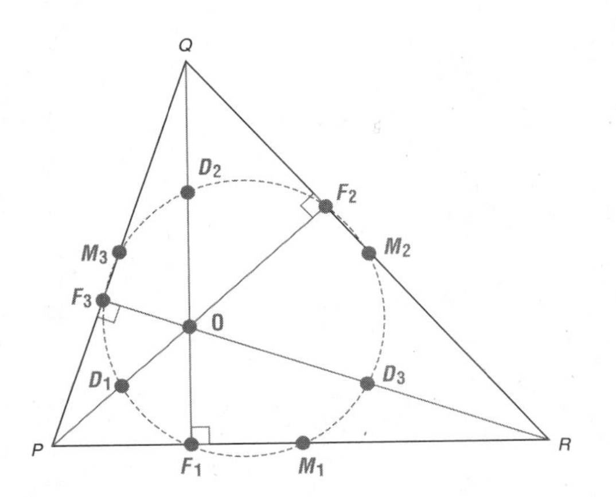

LESSON 8-3
Problem Solving
Circles

Round to the nearest tenth. Use 3.14 for π. Write the correct answer.

1. The world's tallest Ferris wheel is in Osaka, Japan, and stands 369 feet tall. Its wheel has a diameter of 328 feet. Find the circumference of the Ferris wheel.

1029.9 ft

2. A dog is on a 15-foot chain that is anchored to the ground. How much area can the dog cover while he is on the chain?

706.5 ft^2

3. A small pizza has a diameter of 10 inches, and a medium has a diameter of 12 inches. How much more pizza do you get with the medium pizza?

34.5 in^2

4. How much more crust do you get with a medium pizza with a diameter of 12 inches than a small pizza with a 10 inch diameter?

6.3 in.

Round to the nearest tenth. Use 3.14 for π. Choose the letter for the best answer.

5. The wrestling mat for college NCAA competition has a wrestling circle with a diameter of 32 feet, while a high school mat has a diameter of 28 feet. How much more area is there in a college wrestling mat than a high school mat?
   - **A** 12.6 ft^2
   - **(B)** 188.4 ft^2
   - **C** 234.8 ft^2
   - **D** 753.6 ft^2

6. Many tire manufacturers guarantee their tires for 50,000 miles. If a tire has a 16-inch radius, how many revolutions of the tire are guaranteed? There are 63,360 inches in a mile. Round to the nearest revolution.
   - **F** 630.6 revolutions
   - **G** 3125 revolutions
   - **(H)** 31,528,662 revolutions
   - **J** 500,000,000 revolutions

7. In men's Olympic discus throwing competition, an athlete throws a discus with a diameter of 8.625 inches. What is the circumference of the discus?
   - **A** 13.5 in.
   - **(B)** 27.1 in.
   - **C** 58.4 in.
   - **D** 233.6 in.

8. An athlete in a discus competition throws from a circle that is approximately 8.2 feet in diameter. What is the area of the discus throwing circle?
   - **(F)** 52.8 ft^2
   - **G** 25.7 ft^2
   - **H** 12.9 ft^2
   - **J** 211.1 ft^2

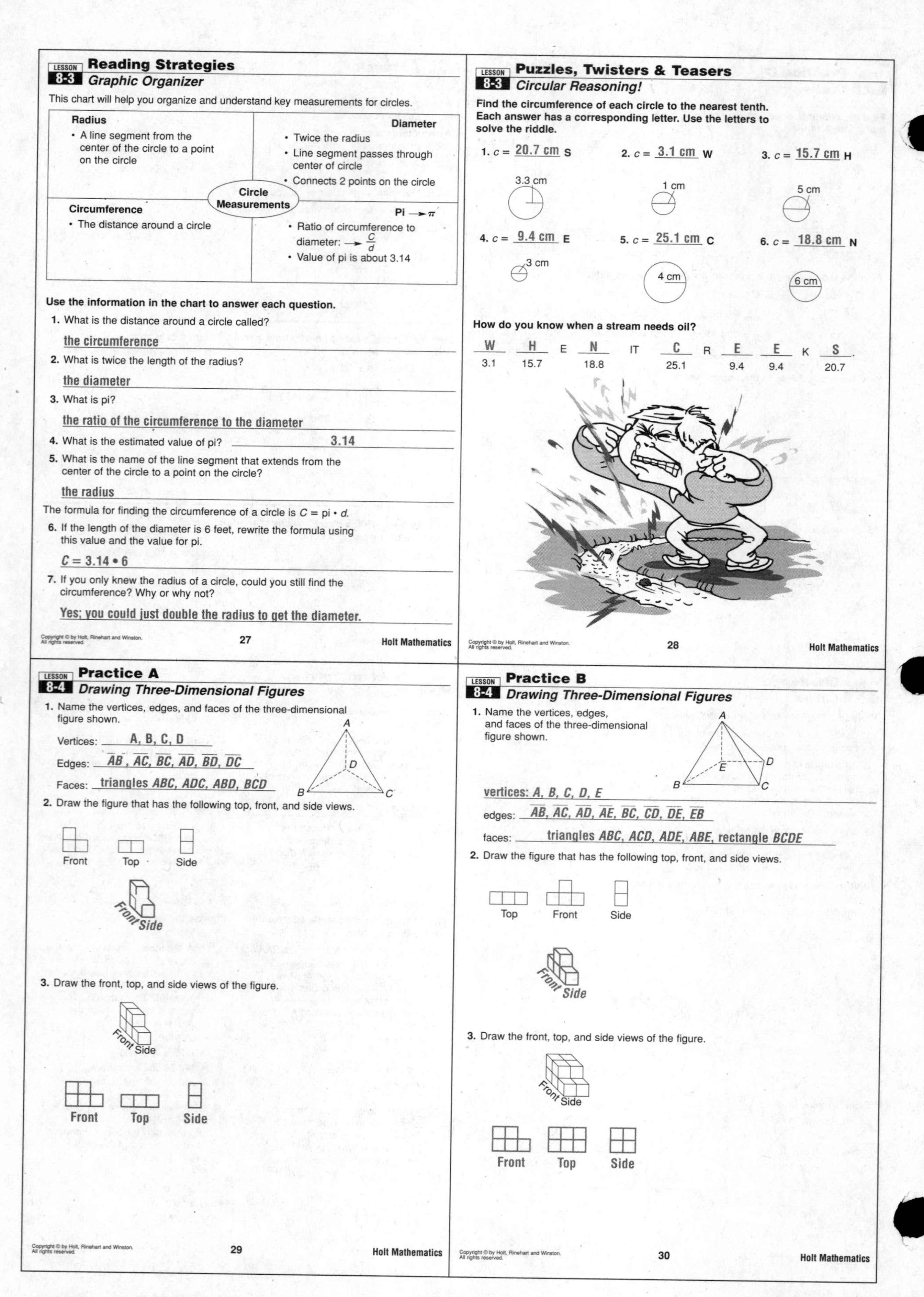

LESSON 8-3

Reading Strategies

Graphic Organizer

This chart will help you organize and understand key measurements for circles.

Radius • A line segment from the center of the circle to a point on the circle	**Diameter** • Twice the radius • Line segment passes through center of circle • Connects 2 points on the circle
Circumference • The distance around a circle	**Pi** → π • Ratio of circumference to diameter: → $\frac{C}{d}$ • Value of pi is about 3.14

Circle Measurements

Use the information in the chart to answer each question.

1. What is the distance around a circle called?

   the circumference

2. What is twice the length of the radius?

   the diameter

3. What is pi?

   the ratio of the circumference to the diameter

4. What is the estimated value of pi? 3.14

5. What is the name of the line segment that extends from the center of the circle to a point on the circle?

   the radius

The formula for finding the circumference of a circle is C = pi • d.

6. If the length of the diameter is 6 feet, rewrite the formula using this value and the value for pi.

   $C = 3.14 \cdot 6$

7. If you only knew the radius of a circle, could you still find the circumference? Why or why not?

   Yes; you could just double the radius to get the diameter.

LESSON 8-3

Puzzles, Twisters & Teasers

Circular Reasoning!

Find the circumference of each circle to the nearest tenth. Each answer has a corresponding letter. Use the letters to solve the riddle.

1. c = 20.7 cm S (3.3 cm)
2. c = 3.1 cm W (1 cm)
3. c = 15.7 cm H (5 cm)
4. c = 9.4 cm E (3 cm)
5. c = 25.1 cm C (4 cm)
6. c = 18.8 cm N (6 cm)

How do you know when a stream needs oil?

W H E N IT C R E E K S.

3.1 15.7 18.8 25.1 9.4 9.4 20.7

LESSON 8-4

Practice A

Drawing Three-Dimensional Figures

1. Name the vertices, edges, and faces of the three-dimensional figure shown.

   Vertices: A, B, C, D

   Edges: $\overline{AB}$, $\overline{AC}$, $\overline{BC}$, $\overline{AD}$, $\overline{BD}$, $\overline{DC}$

   Faces: triangles ABC, ADC, ABD, BCD

2. Draw the figure that has the following top, front, and side views.

   Front Top Side

   Front Side

3. Draw the front, top, and side views of the figure.

   Front Side

   Front Top Side

LESSON 8-4

Practice B

Drawing Three-Dimensional Figures

1. Name the vertices, edges, and faces of the three-dimensional figure shown.

   vertices: A, B, C, D, E

   edges: $\overline{AB}$, $\overline{AC}$, $\overline{AD}$, $\overline{AE}$, $\overline{BC}$, $\overline{CD}$, $\overline{DE}$, $\overline{EB}$

   faces: triangles ABC, ACD, ADE, ABE, rectangle $BCDE$

2. Draw the figure that has the following top, front, and side views.

   Top Front Side

   Front Side

3. Draw the front, top, and side views of the figure.

   Front Side

   Front Top Side

LESSON **8-4** **Practice C**

Drawing Three-Dimensional Figures

Give the dimensions of the base and height in each figure. Name the faces that are parallel.

1. Draw the figure that has the following top, front, and side views.

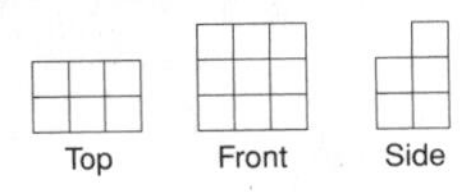

2. Draw the front, top, and side views of the figure.

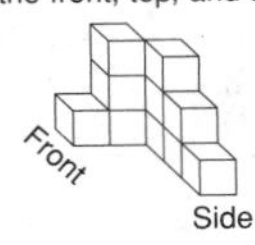

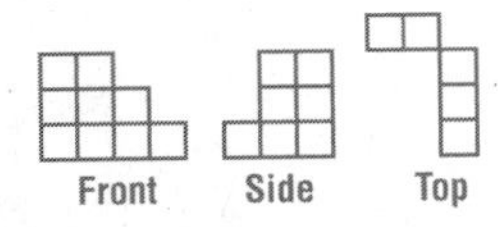

Use isometric dot paper to sketch each figure.

3. a triangular box 4 units high
sample answer:

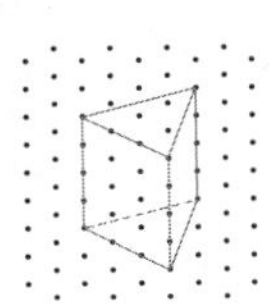

4. a rectangular box 2 units high with a base 4 units by 2 units **sample answer:**

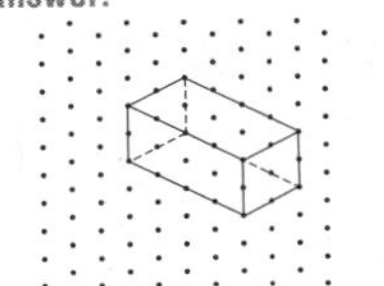

LESSON **8-4** **Reteach**

Drawing Three-Dimensional Figures

Some elements of a three-dimensional figure are:
face, a flat surface that is a plane figure
edge, the line where two faces meet
vertex, the point where three or more edges meet. → Use **vertices** for more than one vertex.

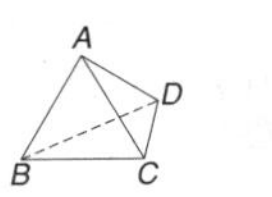

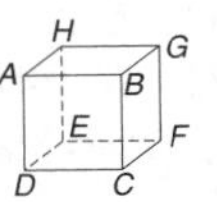

The triangular pyramid has 4 vertices: *A, B, C, D*.
The faces of the figure are triangles. The figure has 4 faces: triangles *ABC, ADC, ABD, BDC*.
The figure has 6 edges.
Trace the lines where the faces meet to name the edges: $\overline{AB}$, $\overline{AC}$, $\overline{AD}$, $\overline{BC}$, $\overline{BD}$, $\overline{CD}$.

The cube has 8 vertices: *A, B, C, D, E, F, G, H*.
The faces of the figure are squares.
The cube has 6 faces: squares *ABCD, ABGH, AHED, EDCF, BCFG, EFGH*.
The cube has 12 edges: $\overline{AB}$, $\overline{BC}$, $\overline{CD}$, $\overline{AD}$, $\overline{AH}$, $\overline{BG}$, $\overline{CF}$, $\overline{DB}$, $\overline{EF}$, $\overline{FG}$, $\overline{GH}$, $\overline{HE}$.

Complete to name the vertices, faces, and edges of the figures shown.

1.

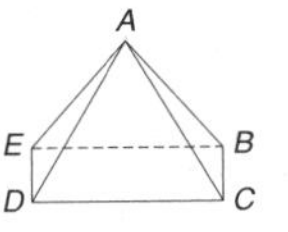

Vertices: *A, B,* __C__, __D__, __E__

Faces: triangles *ABC, ACD,* __ABE__, __AED__, rectangle __BCDE__

Edges: $\overline{AB}$, $\overline{AC}$, $\overline{AD}$, $\overline{AE}$, __$\overline{BC}$__, __$\overline{CD}$__, __$\overline{DE}$__, __$\overline{EB}$__

2.

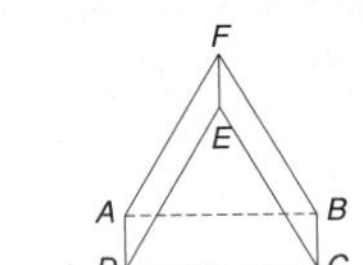

Vertices: *A, B,* __C__, __D__, __E__, __F__

Faces: triangles *ABF,* __CDE__, rectangles ABCD, __BCEF__, __ADEF__

Edges: $\overline{AB}$, $\overline{BC}$, $\overline{CD}$, $\overline{AD}$, __$\overline{AF}$__, __$\overline{BF}$__, __$\overline{FE}$__, __$\overline{DE}$__, __$\overline{CE}$__

LESSON **8-4** **Reteach**

Drawing Three-Dimensional Figures (continued)

You can use the **front**, **side**, and **top** views of a three-dimensional figure to show how the figure looks from different perspectives. When the figures are constructed from cubes, the views will be different groups of squares.

From the front, the stack of cubes appears to have
1 cube on top,
2 cubes in the center, and
3 cubes on the bottom.
The figure looks like 1 square on top of 2 squares that are on top of 3 squares.

From the side, there appears to be a stack 3 cubes high.
The figure looks like 3 squares tall.

From the top, there appears to be a row 3 cubes long.
The figure looks like 3 squares long.

Front View

Side View

Top View

Complete to draw the front, side, and top view of the figure shown.

3.

From the front, the stack of cubes appears to have __2__ cubes on top, __3__ cubes in the center, and __3__ cubes on the bottom.

From the side, there appears to be a stack __3__ cubes high.

From the top, there appears to be a row __3__ cubes long.
The figure looks like 3 squares long.

Front View

Side View

Top View

LESSON **8-4** **Challenge**

From the Ground, Up!

Based on an *isometric drawing* of a solid figure, a **foundation drawing** shows only the base of the solid figure — as a pattern of squares, with a number in each square to tell how many cubes are above it.

Isometric Drawing

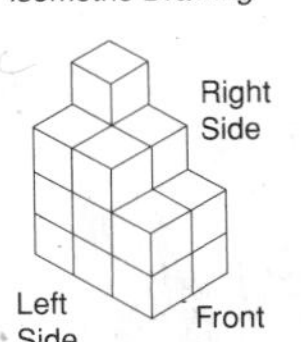

Foundation Drawing

Left Side / Right Side

3	4
3	3
2	2

Front

The front row of the figure is just two cubes high. So, there is a 2 in each box in the front row of the foundation drawing.

The right rear corner column, the highest in the figure, is four cubes high. So, there is a corresponding 4 in the foundation drawing.

Make a foundation drawing for each solid figure.

1.

3	3
3	3
2	2

2.

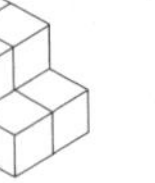

2	2
2	2
1	1

3.

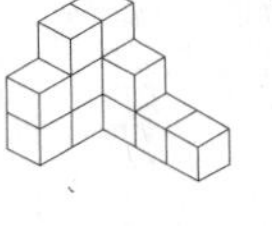

2	3	3
		2
		1
		1

4.

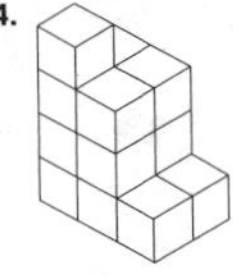

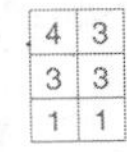

4	3
3	3
1	1

LESSON 8-4
Problem Solving
Drawing Three-Dimensional Figures

Write the correct answer.

1. Mitch used a triangular prism in his science experiment. Name the vertices, edges, and faces of the triangular prism shown at the right.

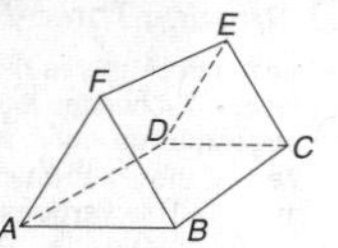

Vertices: *A, B, C, D, E, F*

Faces: triangles *ABF, CDE*, rectangles *ABCD, BCEF, ADEF*

Edges: $\overline{AB}$, $\overline{BC}$, $\overline{CD}$, $\overline{AD}$, $\overline{AF}$, $\overline{BF}$, $\overline{FE}$, $\overline{DE}$, $\overline{CE}$

2. Amber used cubes to make the model shown below of a sculpture she wants to make. Draw the front, side, and top views of the model.

Front

Side

Top

Choose the letter of the best answer.

3. Which is the front view of the figure shown at the left? B

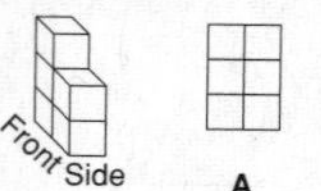

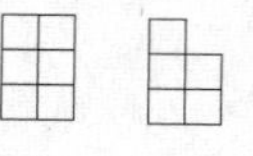

A B C D

4. Which is **not** an edge of the figure shown at the right?
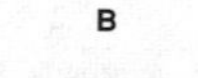

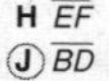

F $\overline{AB}$ H $\overline{EF}$
G $\overline{EH}$ (J) $\overline{BD}$

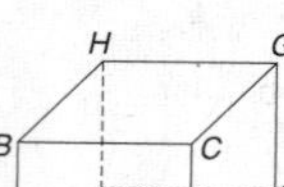
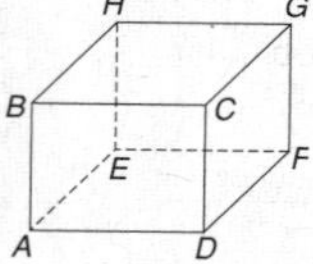

LESSON 8-4
Reading Strategies
Using a Model

This cube is made up of faces, edges, and vertices.

- A **face** is a flat surface.
- An **edge** is where two faces meet.
- A **vertex** is where three or more edges meet.

vertex edge face

When you draw a cube on a flat surface, it is impossible to see all six faces.

A **net** is like a wrapper for the cube. When it is unwrapped and lying flat, you can see all the faces.

Every solid figure has a net.

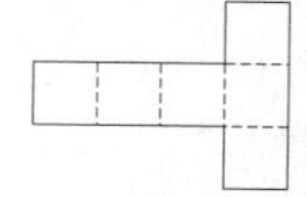

For Exercises 1–4, match each solid figure with its net. Write a, b, c, or d.

a.
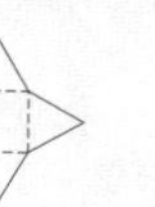

b.

c.
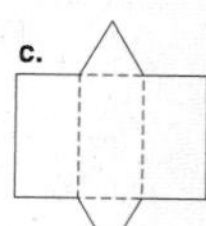

d.
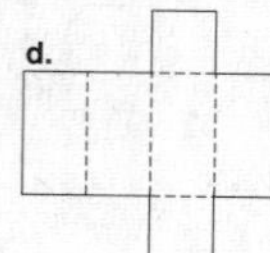

1.
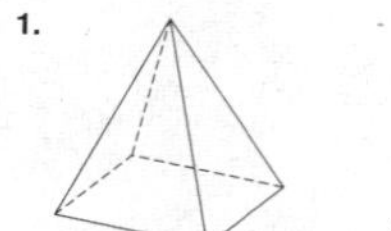

a

2.
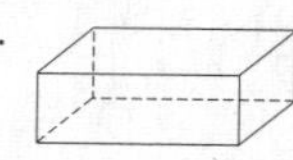
d

3.
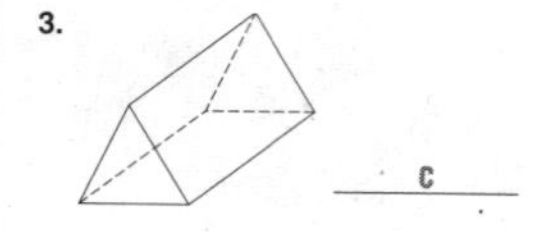

c

4.
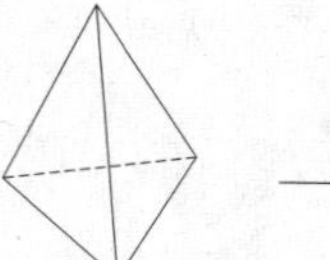
b

LESSON 8-4
Puzzles, Twisters & Teasers
A Three-Dimensional State!

Circle words from the list in the word search (horizontally, vertically or diagonally). Then find a word that answers the riddle. Circle it and write it on the line.

face	edge	vertex	vertices	top	front
side	dimension	view	cube	perspective	orthogonal

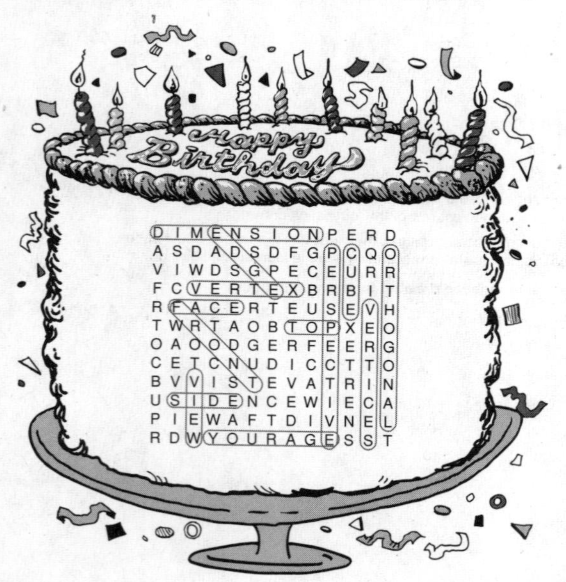

What goes up but never comes down? YOUR AGE

LESSON 8-5
Practice A
Volume of Prisms and Cylinders

Find the volume to the nearest tenth of a unit. Prism: $V = Bh$. Cylinder: $V = \pi r^2 h$. Use 3.14 for π.

1. 3 ft, 5 ft, 12 ft

180 ft^3

2.
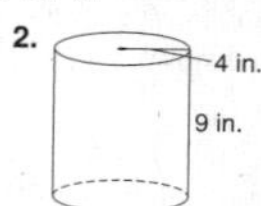

452.2 in^3

3.
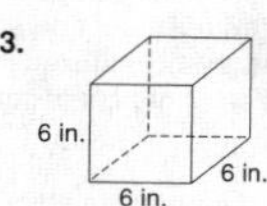

216 in^3

4.
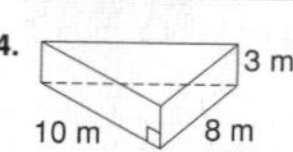

120 m^3

5.
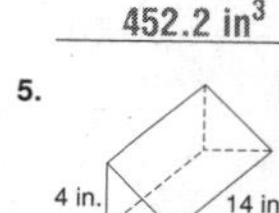

112 in^3

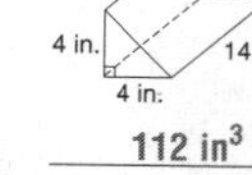

6.
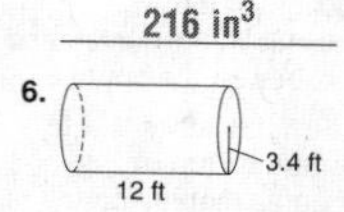

435.6 ft^3

7.
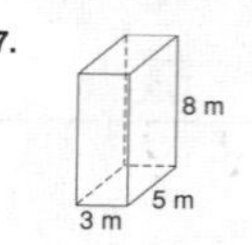

120 m^3

8. 5 cm, 15 cm, 5 cm

187.5 cm^3

9.
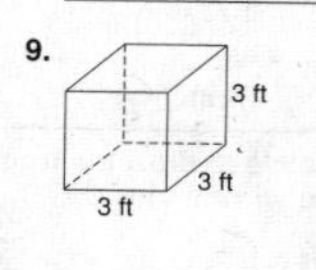

27 ft^3

10. A rectangular box measures 6 ft by 8 ft by 2 ft. Explain whether doubling a side from 6 ft to 12 ft would double the volume of the box. Possible answer: The original box has a volume of 96 ft^3. You could double the volume to 192 ft^3 by doubling any one of the dimensions.

11. A can of vegetables is 4.5 in. high and has a diameter of 3 in. Find the volume of the can to the nearest tenth of a unit. Use 3.14 for π. 31.8 in^3

12. A telephone pole is 30 ft tall with a diameter of 12 in. Jacob is making a replica of a telephone pole and wants to fill it with sand to help it stand freely. Find the volume of his model, which has a height of 30 in. and a diameter of 1 in., to the nearest tenth of a unit. Use 3.14 for π. 23.6 in^3

LESSON 8-5
Practice B
Volume of Prisms and Cylinders

Find the volume of each figure to the nearest tenth. Use 3.14 for π.

1. 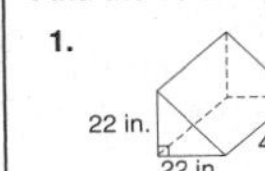

10,164 in^3

2. 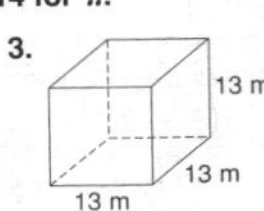

2122.6 cm^3

3. 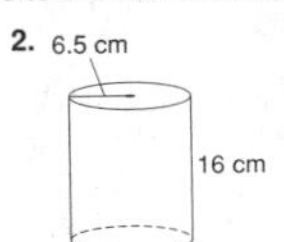

2197 m^3

4. 

3240 cm^3

5. 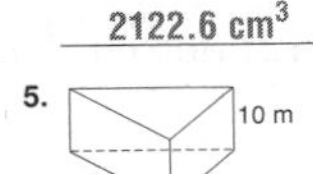

2520 m^3

6. 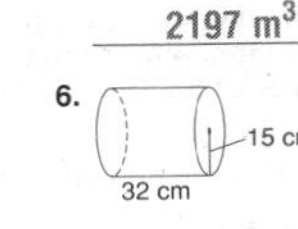

22,608 cm^3

7. 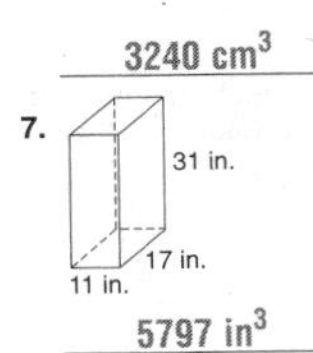

5797 in^3

8. 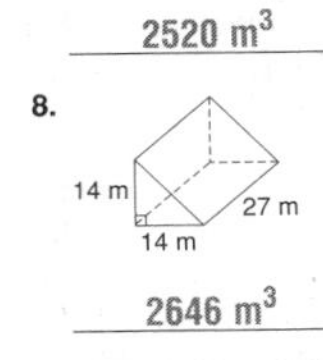

2646 m^3

9. 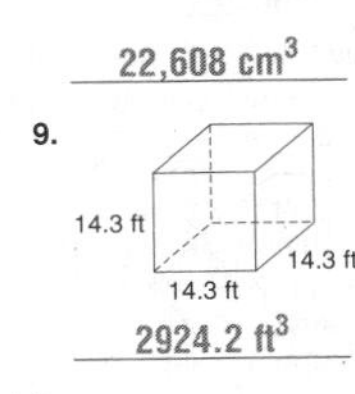

2924.2 ft^3

10. A cylinder has a radius of 6 ft and a height of 25 ft. Explain whether tripling the height will triple the volume of the cylinder. Possible answer: The original cylinder has a volume of 2826 ft^3. If you triple the height the volume is 8478 ft^3, which is triple the original volume.

11. Contemporary American building bricks are rectangular blocks with the standard dimensions of about 5.7 cm by 9.5 cm by 20.3 cm. What is the volume of a brick to the nearest tenth of a unit?
1099.2 cm^3

12. Ian is making candles. His cylindrical mold is 8 in. tall and has a base with a diameter of 3 in. Find the volume of a finished candle to the nearest tenth of a unit.
56.5 in^3

LESSON 8-5
Practice C
Volume of Prisms and Cylinders

Find the volume of each figure to the nearest tenth. Use 3.14 for π.

1. 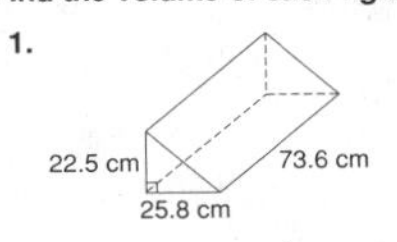

21,362.4 cm^3

2. 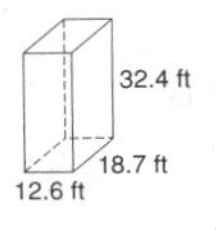

7634.1 ft^3

3. 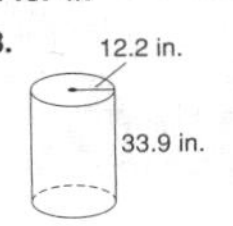

15,843.4 in^3

Find the missing measure to the nearest tenth. Use 3.14 for π.

4. volume is 1152 ft^3

12 ft

5. volume is 1132.8 cm^3

23.6 cm

6. volume is 39,193.48 in^3
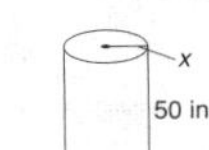

15.8 in.

7. Atmo's Theater's large-sized popcorn container is a cylinder with a diameter of 9 in. and is 6 in. tall. The medium-sized container has a diameter of 6 in. and is 8 in. tall. Find the difference in the volumes. 155.43 in^3

8. The Blackmans have a new 10-ft by 20-ft rectangular swimming pool in their backyard. It has a uniform depth of 5 ft. What is the volume of water needed to fill the pool? 1000 ft^3

9. Donna decorated a three-tiered cake for her parent's anniversary party. The bottom tier had a 12-in. diameter, the middle tier, a 9-in. diameter, and the top tier, a 6-in. diameter. Each tier was 4 in. tall. What was the total volume of the cake to the nearest tenth of a unit? 819.5 in^3

10. A company produces 100,000 boxes of cereal with dimensions of 12 in. (height) by 8 in. by 1.5 in. The manufacturer is trying to cut production cost per box so the height of the box is decreased by one inch. Approximately how many more boxes of cereal can be produced in a month with the new dimensions using the same amount of cereal? Round the answer to the nearest integer. 9091 boxes

/10

LESSON 8-5
Reteach
Volume of Prisms and Cylinders

Volume = number of cubic units inside a solid figure

To find the volume of this solid figure:

Count the number of cubic centimeters in one "slice" of the figure.
$4 \times 3 = 12$

Multiply by the number of "slices." $12 \times 6 = 72$ cm^3

Complete to find the volume of each solid figure.

1. 5

number in^3 in a slice = 4 × 2 = 8
number of slices = 2
volume = 16 in^3

2. 5

number cm^3 in a slice = 3 × 5 = 15
number of slices = 4
volume = 60 cm^3

3. 6

number mm^3 in a slice = 5 × 6 = 30
number of slices = 4
volume = 120 mm^3

/15

LESSON 8-5
Reteach
Volume of Prisms and Cylinders (continued)

Prism: solid figure named for the shape of its two congruent bases

Volume V of a prism = area of base B × height h

$V = Bh$
$V = (6 \times 4)5$
$V = 24(5)$
$V = 120$ in^3

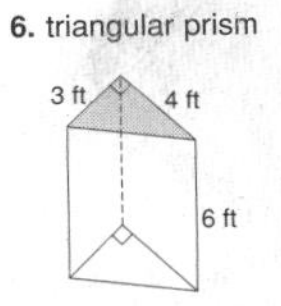

Rectangular Prism

Complete to find the volume of each prism.

4. rectangular prism 4

base is a rectangle
$V = Bh$
$V = (8 \times 6) \times 4$
= 192 cm^3

5. cube 5
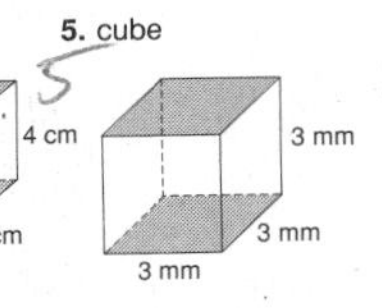

base is a square
$V = Bh$
$V = (3 \times 3) \times 3$
= 27 mm^3

6. triangular prism 5
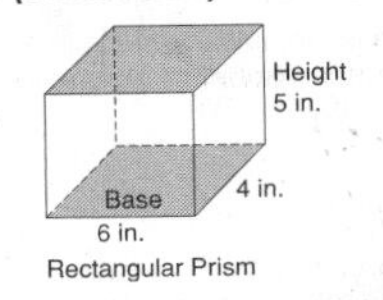

base is a right △
$V = \frac{1}{2} Bh$
$V = \frac{1}{2}(3 \times 4)$
× 6 = 36 ft^3

Cylinder: solid figure with a circular base

Volume V of a cylinder = area of base B × height h

$V = Bh$
$V = (\pi \times 4^2)7 = (16\pi)7$
$V = 112\pi$
$V \approx 112(3.14) \approx 351.7$ $units^3$

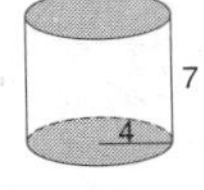
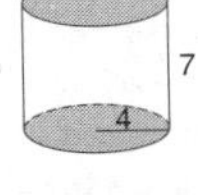

Complete to find the volume of the cylinder.

7. 4
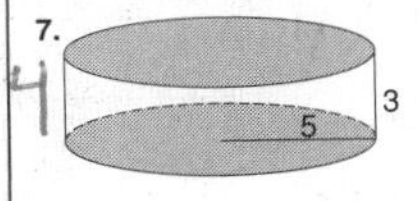

$V = Bh$
$V = (\pi \times 5^2) \times 3$
$V = 75\pi \approx 235.5$ $units^3$

/18

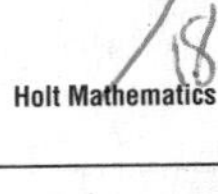
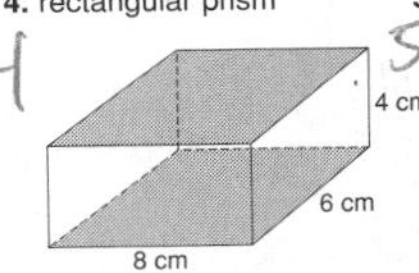
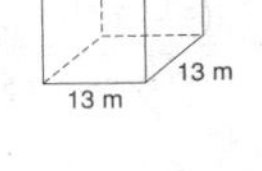

LESSON 8-5
Challenge
Looking Askance

So far, you have considered prisms in which the outside edges are perpendicular to the plane of the base.

Now, you will consider prisms in which the outside edges are not perpendicular to the plane of the base.

Right Prism   Oblique Prism

1. Explain why these two prisms have the same volume.

Areas of bases are equal; heights are equal.

A prism can have any polygon as its base. Consider an oblique prism with a base that is a parallelogram.

To find the volume of this prism, first look at parallelogram *JKLM* which is the base of the prism.

2. How long is *JK*, the base of parallelogram *JKLM*?

$JK =$ 9 units

3. Find *MQ*, the height of parallelogram *JKLM*. Explain your method.

$MQ =$ 4 units; Pythagorean Theorem

4. What is the area of the base of the prism?

$B =$ $9 \times 4 = 36$ units^2

5. Find *KR*, the height of the prism. Explain your method.

$KR =$ 12 units; Pythagorean Theorem

6. Find the volume of the prism.

$V =$ $36 \times 12 = 432$ units^3

LESSON 8-5
Problem Solving
Volume of Prisms and Cylinders

Round to the nearest tenth. Write the correct answer.

1. A contractor pours a sidewalk that is 4 inches deep, 1 yard wide, and 20 yards long. How many cubic yards of concrete will be needed? (Hint: 36 inches = 1 yard.)

2.2 yd^3

2. A refrigerator has inside measurements of 50 cm by 118 cm by 44 cm. What is the capacity of the refrigerator?

259,600 cm^3

A rectangular box is 2 inches high, 3.5 inches wide and 4 inches long. A cylindrical box is 3.5 inches high and has a diameter of 3.2 inches. Use 3.14 for π. Round to the nearest tenth.

3. Which box has a larger volume?

Cylinder

4. How much bigger is the larger box?

0.1 in^3

Use 3.14 for π. Choose the letter for the best answer.

5. A child's wading pool has a diameter of 5 feet and a height of 1 foot. How much water would it take to fill the pool? Round to the nearest gallon. (Hint: 1 cubic foot of water is approximately 7.5 gallons.)
   - A 79 gallons
   - B 589 gallons
   - C 59 gallons
   - **D** 147 gallons

6. How many cubic feet of air are in a room that is 15 feet long, 10 feet wide and 8 feet high?
   - F 33 ft^3
   - **G** 1200 ft^3
   - H 1500 ft^3
   - J 3768 ft^3

7. How many gallons of water will the water trough hold? Round to the nearest gallon. (Hint: 1 cubic foot of water is approximately 7.5 gallons.)

2 ft
6 ft

   - A 19 gallons
   - **B** 71 gallons
   - C 141 gallons
   - D 565 gallons

8. A can has diameter of 9.8 cm and is 13.2 cm tall. What is the capacity of the can? Round to the nearest tenth.
   - F 203.1 cm^3
   - **G** 995.2 cm^3
   - H 3980.7 cm^3
   - J 959.2 cm^3

LESSON 8-5
Reading Strategies
Using a Model

Volume is the amount of space a solid figure occupies. Volume is measured in cubic units.

One cubic unit

A **rectangular prism** is a three-dimensional figure. The length, width, and height are the measures that make up the three dimensions.

4   2   3

You can find the volume of a rectangular prism by multiplying the length times the width times the height.

The formula for the volume of a rectangular prism is:
$V = \ell \cdot w \cdot h$.

Use the rectangular prism to complete Exercises 1–6.

1. How long is the prism? 3 units
2. How wide is the prism? 2 units
3. What is the height of the prism? 3 units
4. Write the formula you will use to find the volume of this prism?

$V = \ell \bullet w \bullet h$

5. Rewrite the formula with values for each dimension of the prism.

$V = 3 \bullet 2 \bullet 3$

6. What is the volume of this prism?

18 cubic units

LESSON 8-5
Puzzles, Twisters & Teasers
Turn Up the Volume!

Across

2. A ________ is a three-dimensional figure named for the shape of its bases.
4. The volume of a prism or cylinder is expressed in ________ units.
6. The volume of a ________ prism can be written as $V = \ell wh$.
9. The volume of a prism is the area of the ________ times the height.

Down

1. The ________ of a three-dimensional figure is the number of cubic units needed to fill it.
3. The area of the base of a prism or cylinder is expressed in ________ units.
5. The volume of a ________ is the area of the base times the height.
7. To find the volume of a ________ three-dimensional figure, find the volume of each part and add the volumes together.
8. If all six faces of a rectangular prism are squares, it is a ________.

Crossword answers: 1 VOLUME, 2 PRISM, 3 SQUARE, 4 CUBIC, 5 CYLINDER, 6 RECTANGULAR, 7 COMPOSITE, 8 CUBE, 9 BASE

LESSON **8-6**

Practice A
Volume of Pyramids and Cones

Find the volume of each figure to the nearest tenth. Use 3.14 for π.

Pyramid: $V = \frac{1}{3}Bh$. Cone: $V = \frac{1}{3}\pi r^2 h$. Use 3.14 for π.

1. 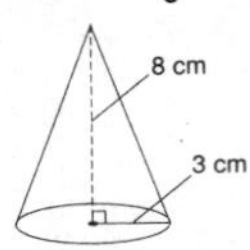

75.4 cm³

2. 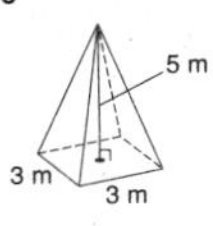

15 m³

3. 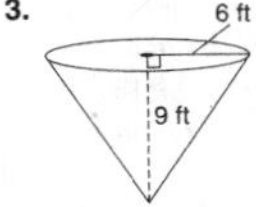

339.1 ft³

4. 6 in., 4 in., 9 in.
72 in³

5. 7.5 ft, 5 ft, 5 ft
62.5 ft³

6. 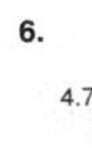
4.7 cm, 8.4 cm
194.2 cm³

7. The base of a regular pyramid has an area of 12 ft². The height of the pyramid is 8 ft. Find the volume. **32 ft³**
8. The radius of a cone is 4.9 cm and its height is 10 cm. Find the volume of the cone to the nearest tenth. Use 3.14 for π. **251.3 cm³**
9. Find the volume of a rectangular pyramid if the height is 6 m and the base sides are 7 m and 5 m. **70 m³**
10. The mold for an ice cone has a diameter of 4 in. and is 5 in. deep. Use a calculator to find the volume of the ice cone mold to the nearest hundredth. **20.94 in³**
11. A square pyramid has a height 6 cm and a base that measures 5 cm on each side. Explain whether doubling the height would double the volume of the pyramid. **Possible answer:**
The volume of the original pyramid is 50 cm³. The volume of the new pyramid is 100 cm³. Therefore, if the height of the pyramid were doubled, its volume would be doubled.

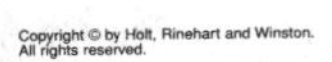

LESSON **8-6**

Practice B
Volume of Pyramids and Cones

Find the volume of each figure to the nearest tenth. Use 3.14 for π.

1. 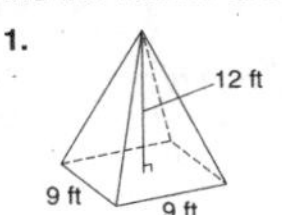

324 ft³

2. 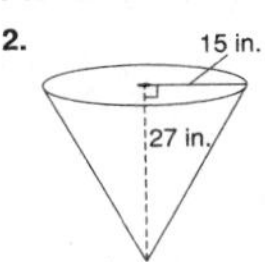

6358.5 in³

3. 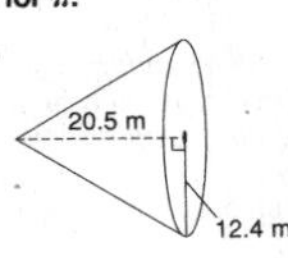

3299.2 m³

4. 23 cm, 19 cm, 20 cm
2913.3 cm³

5. 16 ft, 18 ft, 18 ft
1728 ft³

6. 17 cm, 16 cm
1138.8 cm³

7. The base of a regular pyramid has an area of 28 in². The height of the pyramid is 15 in. Find the volume. **140 in³**
8. The radius of a cone is 19.4 cm and its height is 24 cm. Find the volume of the cone to the nearest tenth. **9454.2 cm³**
9. Find the volume of a rectangular pyramid if the height is 13 m and the base sides are 12 m and 15 m. **780 m³**
10. A funnel has a diameter of 9 in. and is 16 in. deep. Use a calculator to find the volume of the funnel to the nearest hundredth. **339.29 in³**
11. A square pyramid has a height 18 cm and a base that measures 12 cm on each side. Explain whether tripling the height would triple the volume of the pyramid. **Possible answer:**
The volume of the original pyramid is 864 cm³. The volume of the new pyramid is 2592 cm³. Therefore, if the height of the pyramid were tripled, its volume would be tripled.

LESSON **8-6**

Practice C
Volume of Pyramids and Cones

Find the volume of each figure to the nearest tenth. Use 3.14 for π.

1. 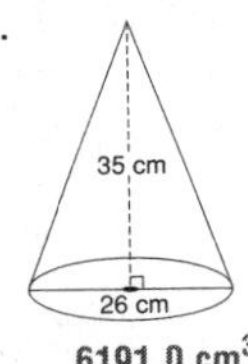

6191.0 cm³

2. 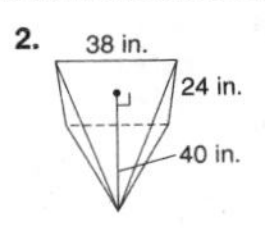

12,160 in³

3. 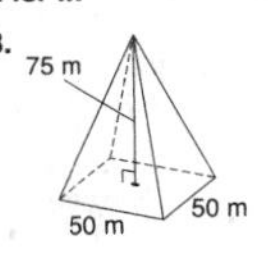

62,500 m³

Find the missing measure to the nearest tenth. Use 3.14 for π.

4. rectangular pyramid:
base length = 15 m
base width = ?
height = 21 m
volume = 2415 m³
base width = 23 m

5. triangular pyramid:
base width = 8 cm
base height = 18 cm
height = ?
volume = 624 cm³
height = 26 cm

6. A cone has diameter of 24 ft and height of 15 ft. How many times will the volume of the cone fill a cylinder with radius of 18 ft and a height of 25 ft? Round your answer to the nearest whole number. **11 times**

7. Find the volume of the figure to the nearest tenth. **3934.9 cm³**
17.5 cm, 14 cm, 16 cm, 12.4 cm

8. Find the volume of the figure to the nearest tenth. **23,314.5 ft³**
21 ft, 26 ft, 15 ft

LESSON **8-6**

Reteach
Volume of Pyramids and Cones

Pyramid: solid figure named for the shape of its base, which is a polygon; all other faces are triangles

Pentagonal Pyramid

This rectangular pyramid and rectangular prism have congruent bases and congruent heights.

Volume of Pyramid = $\frac{1}{3}$ Volume of Prism

$V = \frac{1}{3}Bh$

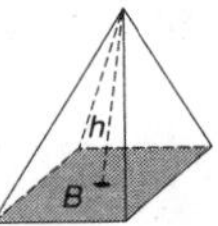

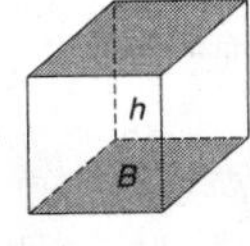

Complete to find the volume of each pyramid.

1. square pyramid

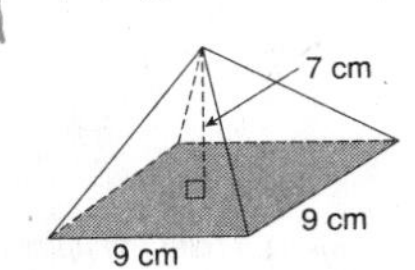

base is a **square**

$V = \frac{1}{3}Bh$

$V = \frac{1}{3}$(area of square) × h

$V = \frac{1}{3}$(**9** × **9**) × **7**

$V = \frac{1}{3}$(**81**) × **7**

$V =$ **189** cm³

2. rectangular pyramid

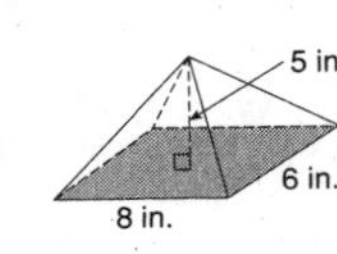
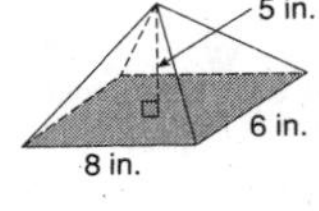

base is a **rectangle**

$V = \frac{1}{3}Bh$

$V = \frac{1}{3}$(area of rectangle) × h

$V = \frac{1}{3}$(**8** × **6**) × **5**

$V = \frac{1}{3}$(**48**) × **5**

$V =$ **80** in³

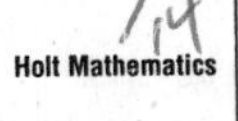

LESSON 8-6 Reteach
Volume of Pyramids and Cones (continued)

Cone: solid figure with a circular base

This cone and cylinder have congruent bases and congruent heights.

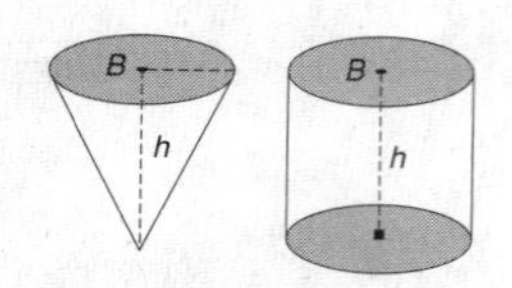

Volume of Cone $= \frac{1}{3}$ Volume of Cylinder

$V = \frac{1}{3}Bh$

Complete to find the volume of each cone.

3.

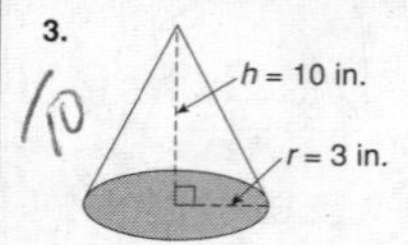

4.

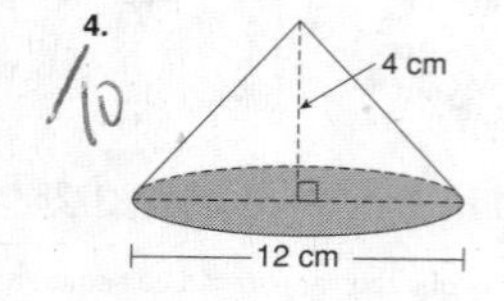
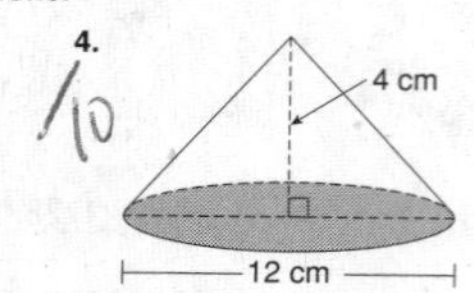

3.	4.
radius r of base = 3 in.	radius $r = \frac{1}{2}$ diameter = 6 cm
$V = \frac{1}{3}Bh$	$V = \frac{1}{3}Bh$
$V = \frac{1}{3}(\pi r^2)h$	$V = \frac{1}{3}(\pi r^2)h$
$V = \frac{1}{3}(\pi \times 3^2) \times 10$	$V = \frac{1}{3}(\pi \times 6^2) \times 4$
$V = \frac{1}{3}(9\pi) \times 10$	$V = \frac{1}{3}(36\pi) \times 4$
$V = 3\pi \times 10$	$V = 12\pi \times 4$
$V = 30\pi$	$V = 48\pi$
$V \approx 30 \times 3.14$	$V \approx 48 \times 3.14$
$V \approx 94.2$ in³	$V \approx 150.72$ cm³

/10 /10 /20

LESSON 8-6 Challenge
Take a Little Off the Top

When a pyramid is cut by a plane parallel to its base, the part of the surface between and including the two planes is called a **truncated pyramid** or **frustum of a pyramid**.

Consider a square pyramid of base 4 cm by 4 cm and height 12 cm. As shown, a plane parallel to the base cuts the pyramid halfway down.

1. Use the familiar formula $V = \frac{1}{3}Bh$ where B = area of base and h = height to find the volume of the large pyramid.

   $V_{\text{large pyramid}} = \frac{1}{3}(16)(12) = 64$ cm³

2. Find the volume of the small pyramid at the top.

   $V_{\text{small pyramid}} = \frac{1}{3}(4)(6) = 8$ cm³

3. Use your results to find the volume.

   $V_{\text{truncated pyramid}} = 64 - 8 = 56$ cm³

A papyrus roll written in ancient Egypt around 1890 B.C.E.—now known as the Moscow Papyrus because it was brought to Russia in 1893—gives a formula for finding the volume of a truncated pyramid.

$V_{\text{truncated pyramid}} = \frac{1}{3}h(a^2 + ab + b^2)$

where
h = height
a = edge length of top base
b = edge length of bottom base

4. Use the ancient formula to find the volume of the truncated pyramid. Compare your results from exercises 3 and 4.

   $V_{\text{truncated pyramid}} = \frac{1}{3}(6)(2^2 + 2(4) + 4^2) = 56$ cm³

/4

LESSON 8-6 Problem Solving
Volume of Pyramids and Cones

Round to the nearest tenth. Use 3.14 for π. Write the correct answer.

1. The Feathered Serpent Pyramid in Teotihuacan, Mexico is the third largest in the city. Its base is a square that measures 65 m on each side. The pyramid is 19.4 m high. What is the volume of the Feathered Serpent Pyramid?

   27,321.7 m³

2. The Sun Pyramid in Teotihuacán, Mexico, is larger than the Feathered Serpent Pyramid. The sides of the square base and the height are each about 3.3 times larger than the Feathered Serpent Pyramid. How many times larger is the volume of the Sun Pyramid than the Feathered Serpent Pyramid?

   35.9 times larger

3. An oil funnel is in the shape of a cone. It has a diameter of 4 inches and a height of 6 inches. If the end of the funnel is plugged, how much oil can the funnel hold before it overflows?

   25.1 in³

4. One quart of oil has a volume of approximately 57.6 in³. Does the oil funnel in exercise 3 hold more or less than 1 quart of oil?

   less

Round to the nearest tenth. Use 3.14 for π. Choose the letter for the best answer.

5. An ice cream cone has a diameter of 4.2 cm and a height of 11.5 cm. What is the volume of the cone?
   - A 18.7 cm³
   - B 25.3 cm³
   - (C) 53.1 cm³
   - D 212.3 cm³

6. When decorating a cake, the frosting is put into a cone shaped bag and then squeezed out a hole at the tip of the cone. How much frosting is in a bag that has a radius of 1.5 inches and a height of 8.5 inches?
   - F 5.0 in³
   - G 13.3 in³
   - H 15.2 in³
   - (J) 20.0 in³

7. What is the volume of the hourglass at the right?
   - (A) 13.1 in³
   - B 26.2 in³
   - C 52.3 in³
   - D 102.8 in³

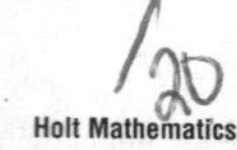

LESSON 8-6 Reading Strategies
Compare and Contrast

Compare the shapes of these three-dimensional figures.

Compare the facts for the cone and the pyramid.

Cone	Pyramid
• three-dimensional figure	• three-dimensional figure
• base is a circle	• base is a polygon
• other face is a curved surface	• other faces are triangles
• one vertex	• at least four vertices

Compare the pictures and the facts about the cone and the pyramid to answer the questions.

1. Which figure has a circular base?

   the cone

2. What is the shape of the base of a pyramid?

   a polygon

3. Which figure has triangles as faces?

   the pyramid

4. How many bases does each figure have?

   one

5. How many dimensions does each figure have?

   three

6. How are the figures different?

   Possible answer: the base of each figure is a different shape.

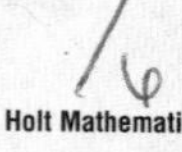

/6

LESSON 8-6

Puzzles, Twisters & Teasers

How Much Volume!

Decide whether or not the volume given for each figure is correct. Circle the letter above your answer. Use the letters to solve the riddle.

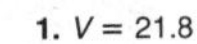

1. $V = 21.8$

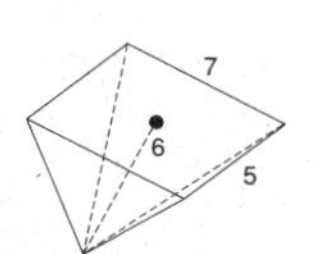

2. $V = 52.5$

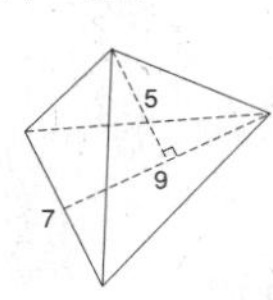

3. $V = 15$

D	(B)	(R)	E	L	(A)
correct	incorrect	correct	incorrect	correct	incorrect

4. $V = 924$

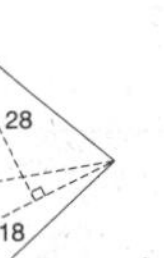

5. $V = 65.56$

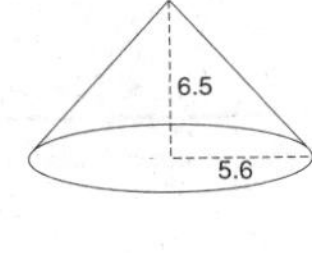

6. $V = 3159$

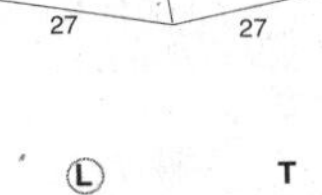

(Z)	R	K	(I)	(L)	T
correct	incorrect	correct	incorrect	correct	incorrect

What kind of nut is like a country?

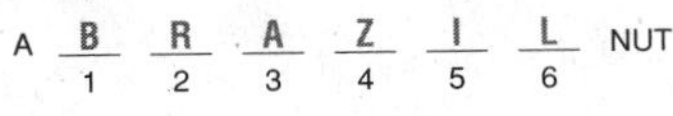

A B R A Z I L NUT
(1 2 3 4 5 6)

LESSON 8-7

Practice A

Surface Area of Prisms and Cylinders

Find the surface area of each figure to the nearest tenth. Prism: $S = 2B + Ph$. Cylinder: $S = 2\pi r^2 + 2\pi rh$. Use 3.14 for π.

1.

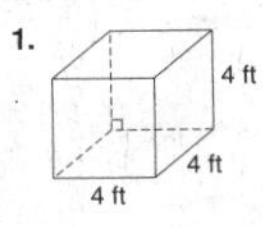

$96\ ft^2$

2. 3 cm, 8 cm

$207.2\ cm^2$

3.

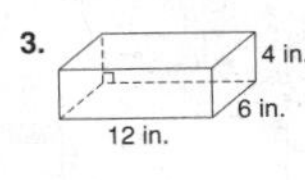

$288\ in^2$

4. 5 m, 12 m, 3 m, 4 m

$156\ m^2$

5.

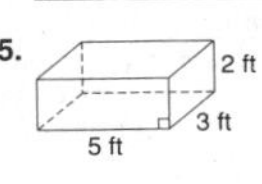
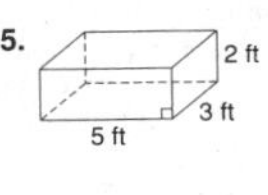

$62\ ft^2$

6. 4 m, 8.5 m

$314\ m^2$

7.

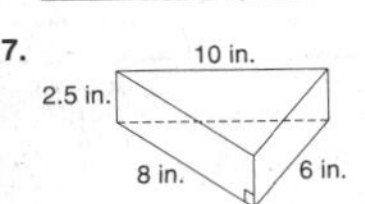

$108\ in^2$

8. 5 ft, 6 ft

$345.4\ ft^2$

9. 9.6 m, 6.8 m, 4.2 m

$268.3\ m^2$

10. Find the surface area to the nearest tenth of a cylinder 50.6 ft tall that has a diameter of 30 ft. Use 3.14 for π. $6179.5\ ft^2$

11. Find the surface area to the nearest tenth of a square prism with sides 6.2 m. $230.6\ m^2$

12. To the nearest tenth, how much paper is needed for the label of a soup can if the can is 8.5 in. tall and has a diameter of 4 in.? Use 3.14 for π. (Hint: The label does not cover the bases of the cylinder.) $106.8\ in^2$

LESSON 8-7

Practice B

Surface Area of Prisms and Cylinders

Find the surface area of each figure to the nearest tenth. Use 3.14 for π.

1.

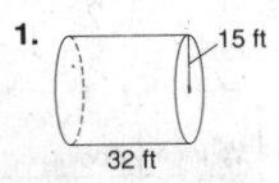

$4427.4\ ft^2$

2.

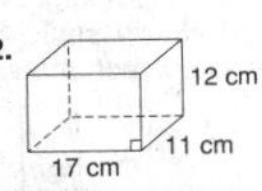

$1046\ cm^2$

3. 15 in., 9 in., 22 in., 12 in.

$900\ in^2$

4. 14 m, 14 m, 14 m

$1176\ m^2$

5. 7.5 cm, 10.5 cm

$847.8\ cm^2$

6.

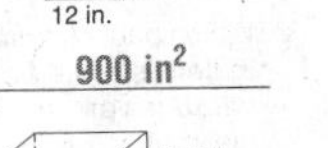
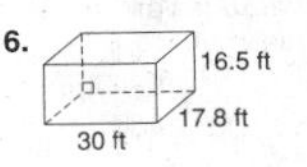

16.5 ft, 17.8 ft, 30 ft

$2645.4\ ft^2$

7. 10 in., 4 in.

$879.2\ in^2$

8. 18.1 ft, 15.3 ft, 12.4 ft

$1382.2\ ft^2$

9. 13 m, 3 m, 12 m, 5 m

$150\ m^2$

10. Find the surface area to the nearest tenth of a rectangular prism with height 15 m and sides 14 m and 13 m. $1174\ m^2$

11. Find the surface area to the nearest tenth of a cylinder 61.7 ft tall that has a diameter of 38 ft. $9629.1\ ft^2$

12. Henry wants to paint the ceiling and walls of his living room. One gallon of paint covers 450 ft^2. The room is 24 ft by 18 ft, and the walls are 9 ft high. How many full gallons of paint will Henry need to paint his living room? 3 gal

13. A rectangular prism is 18 in. by 16 in. by 10 in. Explain the effect, if any, tripling all the dimensions will have on the surface area of the figure. **Possible answer:** By tripling the dimensions, the surface area becomes 9 times larger, from 1256 in^2 to 11,304 in^2.

LESSON 8-7

Practice C

Surface Area of Prisms and Cylinders

Find the surface area of each figure to the nearest tenth. Use 3.14 for π.

1.

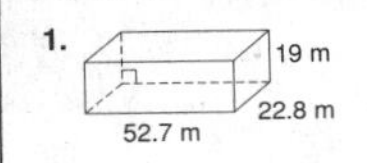

$5272.1\ m^2$

2.

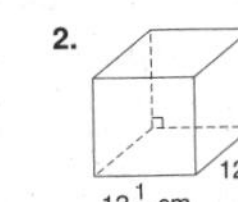

$12\frac{1}{3}$ cm, $12\frac{1}{3}$ cm, $12\frac{1}{3}$ cm

$912\frac{2}{3}\ cm^2$

3. 26 in., 24 in., 51 in.

$3300\ in^2$

4. Germane is making a rectangular cardboard house for his pet gerbil. He wants the house to be at least 5 in. tall, 6 in. wide, and 8 in. long. What is the minimum surface area, if the base of the rectangular prism is open? $188\ in^2$

5. Galena is a common isometric mineral. It occurs naturally in the form of a cube. What is the surface area of a crystal of galena to the nearest tenth if its sides measure 8.4 cm? $423.4\ cm^2$

6. Cristina is painting the playhouse she made for her sister, Delia. She intends to paint only the outside of the playhouse, which does not have a floor. The paint she chose must be purchased in full gallons. Each gallon covers 200 ft^2 and costs $15.95 a gallon. How much will it cost Cristina to paint the playhouse? $31.90

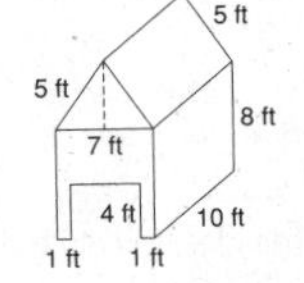

Find the missing dimension in each figure with the given surface area.

7. surface area = 1248 in^2

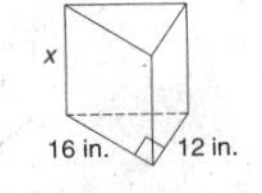

height = 22 in.

8. surface area = 628 ft^2

5 ft, x

height = 15 ft

LESSON 8-7
Reteach
Surface Area of Prisms and Cylinders

Find the number of tiles needed to cover the faces of the prism. Unfold the prism to get a better look at its six faces.

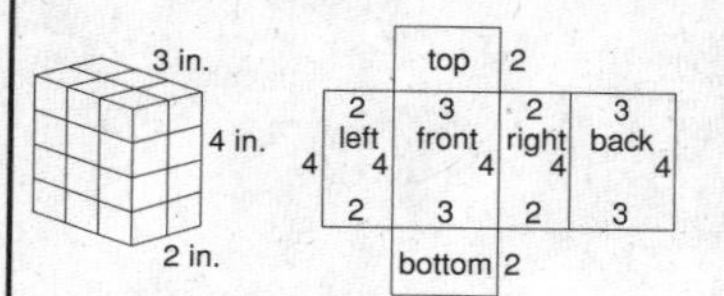

Face	Area	in²
top	3 × 2	6
bottom	3 × 2	6
front	3 × 4	12
back	3 × 4	12
left	2 × 4	8
right	2 × 4	8
	total =	52 tiles

Surface Area S = the sum of the areas of the faces of the prism
= top + bottom + front + back + left + right
= area of bases + area of lateral faces
= $2B$ + perimeter of the base × height of prism

$\mathbf{S = 2B + Ph}$

$S = 2(3 \times 2) + (3 + 2 + 3 + 2) \times 4$

$S = 12 + (10) \times 4$

$S = 12 + 40$

$S = 52 \text{ in}^2$

1. Complete to find the number of square units needed to cover all the faces of the rectangular prism.

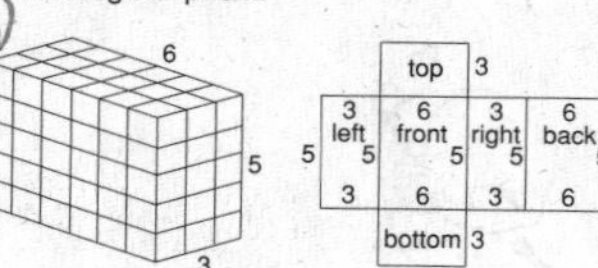

Face	Area	in²
top	6 × 3	18
bottom	6 × 3	18
front	6 × 5	30
back	6 × 5	30
left	3 × 5	15
right	3 × 5	15
	total =	126

2. Complete to find the surface area of the prism.

$S = 2B + Ph$

$S = 2(\underline{5 \times 7}) + (\underline{5 + 7 + 5 + 7}) \times \underline{3}$

$S = 2(\underline{35}) + (\underline{24}) \times \underline{3}$

$S = \underline{70} + \underline{72} = \underline{142} \text{ in}^2$

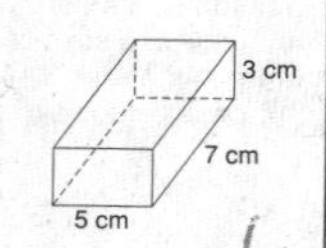

LESSON 8-7
Reteach
Surface Area of Prisms and Cylinders (continued)

An unfolded cylinder results in two circles and a lateral surface drawn as a rectangle.

The base of the rectangle equals the circumference of the circular base.

The height of the rectangle equals the height of the cylinder.

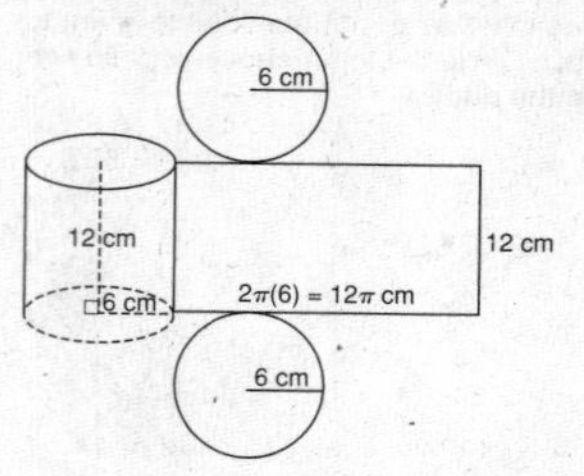

Surface Area S = area of 2 circular bases + area of lateral surface (rectangle)
= $2(\pi r^2)$ + circumference × height
= $2\pi r^2$ + $2\pi r \times h$

$\mathbf{S = 2\pi r^2 + 2\pi rh}$

$S = 2\pi(6^2) + 2\pi(6)(12)$

$S = 72\pi + 144\pi = 216\pi \text{ cm}^2$

$S \approx 216(3.14) \approx 678.24 \text{ cm}^2$

3. Complete to find the surface area of the cylinder.

$S = 2\pi r^2 + 2\pi rh$

$S = 2\pi(\underline{12^2}) + 2\pi \times \underline{12} \times \underline{8}$

$S = \underline{288}\pi + \underline{192}\pi$

$S = \underline{480}\pi$

$S \approx \underline{480}(3.14)$

$S \approx \underline{1507.2} \text{ in}^2$

8 in.
12 in.

Find the surface area of each cylinder. Round to the nearest whole number.

4. height = 10 ft, radius = 5 ft

471 ft^2

5. height = 2.5 cm, diameter = 8 cm

163 cm^2

LESSON 8-7
Challenge
Eight Snips

A **cube** is a prism with six congruent square faces and eight vertices.

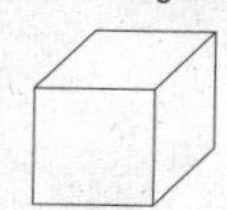

By cutting off the corners of the cube $\frac{1}{3}$ of the way into each edge, a **truncated cube** is created.

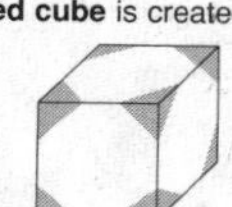

1. Cut along the outer perimeter of this pattern for a truncated cube. Carefully cut into the notched areas found next to some of the shaded sections.

Fold down to make a crease along each line.

Using the shaded sections as "underlaps," tape the figure together.

2. How many faces in all?

14

3. Describe the nature of the faces.

8 equilateral triangles and 6 octagons

4. Find the surface area of a truncated cube. The original cube had edge length = 24 in. Use $\sqrt{3} \approx 1.73$. Answer to the nearest tenth of a square inch.

3130.88 in^2

LESSON 8-7
Problem Solving
Surface Area of Prisms and Cylinders

An important factor in designing packaging for a product is the amount of material required to make the package. Consider the three figures described in the table below. Use 3.14 for π. Round to the nearest tenth. Write the correct answer.

1. Find the surface area of each package given in the table.

2. Which package has the lowest materials cost? Assume all of the packages are made from the same material.

cylinder

Package	Dimensions	Volume	Surface Area
Prism	Base: 2" × 16" Height = 2"	64 in³	136 in²
Prism	Base: 4" × 4" Height = 4"	64 in³	96 in²
Cylinder	Radius = 2" Height = 5.1"	64.06 in³	89.2 in²

Use 3.14 for π. Round to the nearest hundredth.

3. How much cardboard material is required to make a cylindrical oatmeal container that has a diameter of 12.5 cm and a height of 24 cm, assuming there is no overlap? The container will have a plastic lid.

1064.66 cm^2

4. What is the surface area of a rectangular prism that is 5 feet by 6 feet by 10 feet?

280 ft^2

Use 3.14 for π. Round to the nearest tenth. Choose the letter for the best answer.

5. How much metal is required to make the trough pictured below?

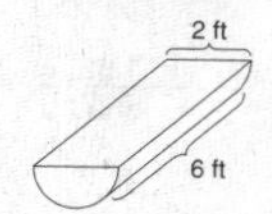

(A) 22.0 ft² (circled)
B 34.0 ft²
C 44.0 ft²
D 56.7 ft²

6. A can of vegetables has a diameter of 9.8 cm and is 13.2 cm tall. How much paper is required to make the label, assuming there is no overlap? Round to the nearest tenth.

F 203.1 cm²
(G) 406.2 cm² (circled)
H 557.0 cm²
J 812.4 cm²

LESSON 8-7

Reading Strategies
Identify Relationships

The **surface area** of a solid figure is the total area of its outside surfaces. You can think of surface area as the part of the solid shape that you can paint. You can paint all the surfaces. Surface area is always given in square units, just like area.

In order to figure out the surface area of a rectangular prism, you can "unfold" it to make a net.

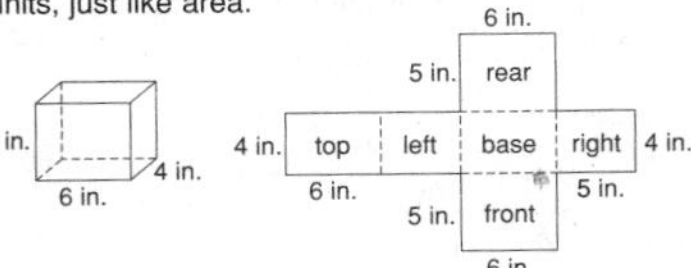

A rectangular prism has six faces, each the shape of a rectangle. To find the surface area of a rectangular prism, find the sum of the areas of the six faces, or rectangles. Opposite faces are equal.

The figure above will help you find the surface area of the rectangular prism. Use the formula $A = (\ell \cdot w)$ to find the area of each face.

Answer each question to find the surface area of the rectangular prism.

1. What is the area of the base rectangle? 24 square inches
2. What is the area of the top rectangle? 24 square inches
3. What is the area of the rear rectangle? 30 square inches
4. What is the area of the front rectangle? 30 square inches
5. What is the area of the right rectangle? 20 square inches
6. What is the area of the left rectangle? 20 square inches

To find the surface area of the rectangular prism, add the area of all six rectangles.

7. What is the surface area of the rectangular prism?

148 square inches

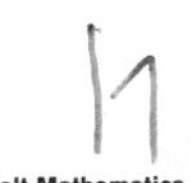

LESSON 8-7

Puzzles, Twisters & Teasers
How Much Volume!

Decide which formula should be used to find the surface area of each figure. Circle the letter above your answer. Then use the letters to solve the riddle.

1. 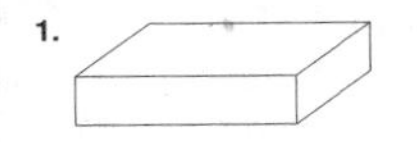

2. 

3. 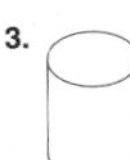 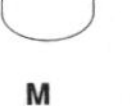

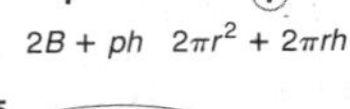

Figure		
1.	(W) $2B + ph$	R $2\pi r^2 + 2\pi rh$
2.	F $2B + ph$	(T) $2\pi r^2 + 2\pi rh$
3.	M $2B + ph$	(R) $2\pi r^2 + 2\pi rh$

4. 

5. 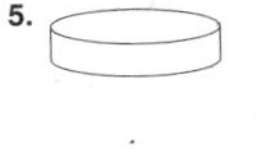

6. 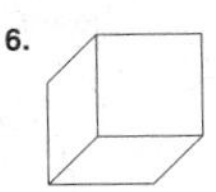

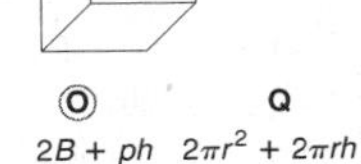

Figure		
4.	(M) $2B + ph$	G $2\pi r^2 + 2\pi rh$
5.	P $2B + ph$	(L) $2\pi r^2 + 2\pi rh$
6.	(O) $2B + ph$	Q $2\pi r^2 + 2\pi rh$

When would you go on red and stop on green?

When you're eating a W(1) A T(2) E R(3) M(4) E L(5) O(6) N.

LESSON 8-8

Practice A
Surface Area of Pyramids and Cones

Find the surface area of each figure to the nearest tenth.
Pyramid: $S = B + \frac{1}{2}P\ell$. Cone: $S = \pi r^2 + \pi r\ell$.
Use 3.14 for π.

1.  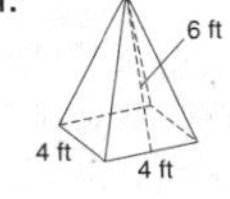

64 ft²

2. 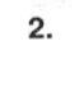 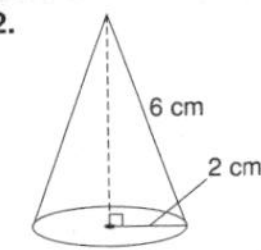

50.2 cm²

3. 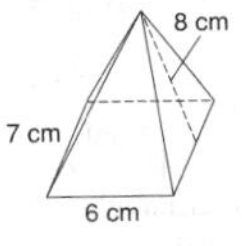

146 cm²

4. 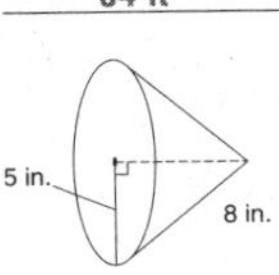

204.1 in²

5. 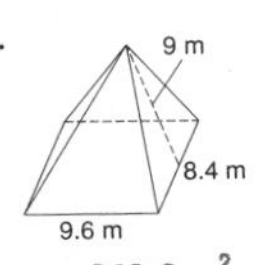

242.6 m²

6. 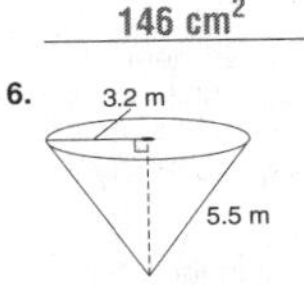

87.4 m²

7. 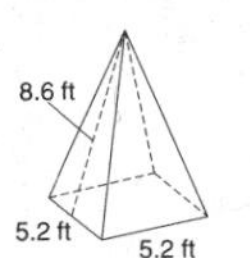

116.5 ft²

8. 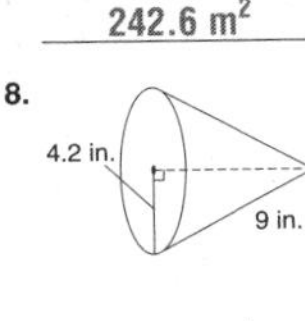

174.1 in²

9. 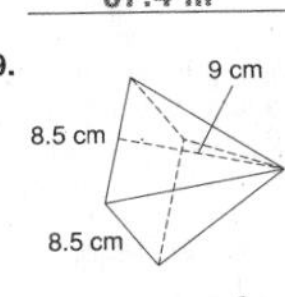

225.3 cm²

10. Find the surface area of a regular square pyramid with a slant height of 7 m and a base perimeter of 10 m.

41.3 m²

11. Find the length of the slant height of a square pyramid if one side of the base is 5 ft and the surface area is 125 ft².

10 ft

12. Find the length of the slant height of a cone with a radius of 5 cm and a surface area of 235.5 cm². Use 3.14 for π.

10 cm

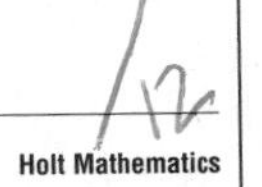

LESSON 8-8

Practice B
Surface Area of Pyramids and Cones

Find the surface area of each figure to the nearest tenth. Use 3.14 for π.

1. 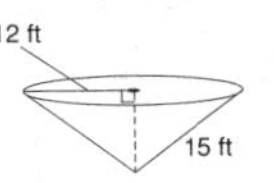

1017.4 ft²

2. 24 ft, 18 ft, 18 ft

1188 ft²

3. 15 cm, 12 cm, 9 cm

423 cm²

4. 13.5 in., 13 in.

1081.7 in²

5. 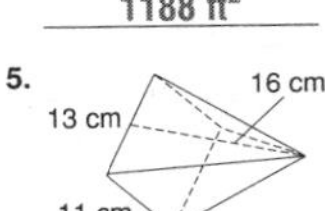

527 cm²

6. 22.5 in., 19.6 in., 19.6 in.

1266.2 in²

7. 18 m, 22 m

2260.8 m²

8. 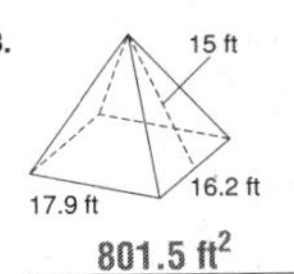

801.5 ft²

9. 15.8 m, 17.6 m

1657.0 m²

10. Find the surface area of a regular square pyramid with a slant height of 17 m and a base perimeter of 44 m. 495 m²

11. Find the length of the slant height of a square pyramid if one side of the base is 15 ft and the surface area is 765 ft². 18 ft

12. Find the length of the slant height of a cone with a radius of 15 cm and a surface area of 1884 cm². 25 cm

13. A cone has a diameter of 12 ft and a slant height of 20 ft. Explain whether tripling both dimensions would triple the surface area. Possible answer:

The surface area of the first cone is 489.84 ft². The surface area of the cone with the new dimensions is 4408.56 ft². It increases the surface area by a factor of 9.

LESSON 8-8
Practice C
Surface Area of Pyramids and Cones

Find the surface area of each figure to the nearest tenth. Use 3.14 for π.

1. 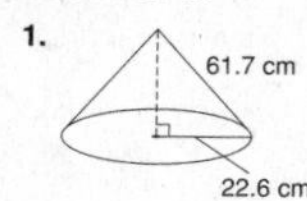

5982.3 cm^2

2. 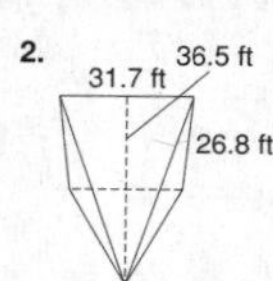

2984.8 ft^2

3. 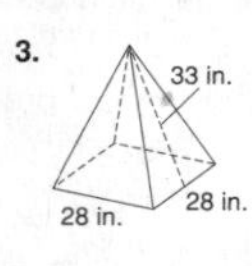

2632 in^2

Find the surface area of each figure with the given dimensions. Use 3.14 for π.

4. regular square pyramid: base perimeter = 80 ft, slant height = 25 ft — 1400 ft^2

5. regular triangular pyramid: base area = 128 in^2, base perimeter = 48 in., slant height = 21 in. — 632 in^2

6. cone: diameter = 31 m, slant height = 37 m — 2555.2 m^2

7. cone: radius = $5\frac{1}{3}$ cm, slant height = 8.5 cm — 231.7 cm^2

8. A cone-shaped roof is attached to a silo. If the diameter of the roof is 25 ft and the slant height is 15 ft, find the surface area of the cone-shaped roof to the nearest tenth. — 588.75 ft^2

9. At 146.5 m high, the Great Pyramid of Egypt stood as the tallest structure in the world for more than 4000 years. It is believed that this great structure was built around 3200 B.C. One base side of the Great Pyramid is approximately 231 m and its slant height is approximately 186 m. Find the approximate surface area of the structure. — $139{,}293 \text{ m}^2$

10. If a model of the Great Pyramid were built with a base side of 7.7 cm and slant height 6.2 cm, what would be the surface area of the model to the nearest tenth? — 154.8 cm^2

10

LESSON 8-8
Reteach
Surface Area of Pyramids and Cones

Regular Pyramid:
base is a regular polygon; lateral faces are congruent triangles

When a square pyramid is unfolded, there are 5 faces: a square and 4 congruent triangles.

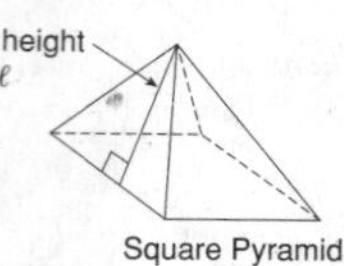

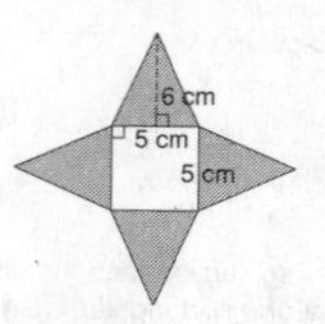

Surface Area S = the sum of the areas of the faces of the pyramid
= area of base + area of lateral faces
Surface Area S = area of square + 4(area of triangle)
= B + $\frac{1}{2}$ perimeter P of base × slant height ℓ of prism

$\mathbf{S = B + \frac{1}{2}P\ell}$
$S = (5 \times 5) + \frac{1}{2} \times (5 \times 4) \times 6$
$S = 25 + 60$
$S = 85 \text{ cm}^2$

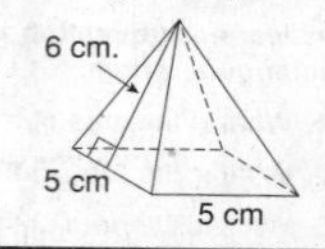

1. Find the surface area of the square pyramid.
S = area of square + 4(area of triangle)

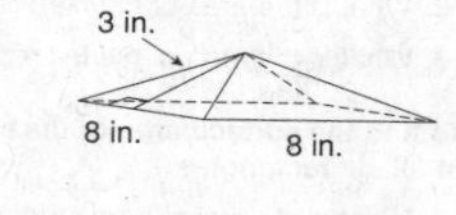

$S = (8 \times 8) + 4(\frac{1}{2} \times 8 \times 3)$
$S = 64 + 48$
$S = 112 \text{ in}^2$

2. Complete to find the surface area of the square pyramid.
$S = B + \frac{1}{2}P\ell$
$S = (9 \times 9) + \frac{1}{2}\ 36 \times 12$
$S = 81 + 216$
$S = 297 \text{ ft}^2$

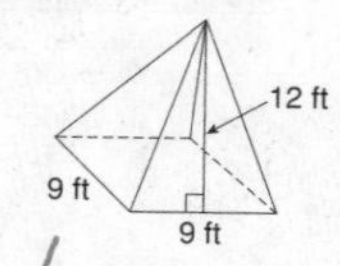

14

LESSON 8-8
Reteach
Surface Area of Pyramids and Cones (continued)

An unfolded cone results in a circle and a lateral surface drawn as a sector of a circle.

The slant height of the cone is the radius of the circle sector.

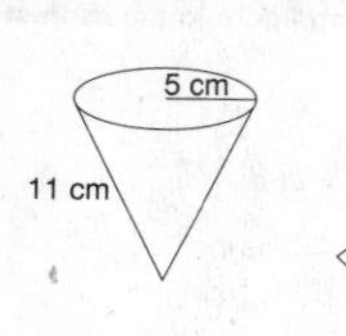

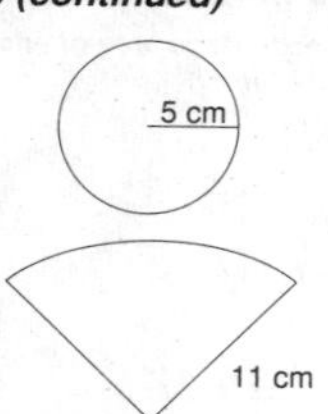

Surface Area S = area of circular base + area of lateral surface (circle sector)
$= \pi r^2 + \frac{1}{2}$ circumference of base × slant height
$= \pi r^2 + \frac{1}{2}(2\pi r) \times \ell$

$\mathbf{S = \pi r^2 + \pi r\ell}$
$S = \pi(5^2) + \pi(5)(11)$
$S = 25\pi + 55\pi = 80\pi \text{ cm}^2$
$S \approx 80(3.14) \approx 251.2 \text{ cm}^2$

Complete to find the surface area of the cone.

3. $S = \pi r^2 + \pi r\ell$
$S = \pi(4^2) + \pi \times 4 \times 7$
$S = 16\pi + 28\pi = 44\pi$
$S \approx 44 (3.14)$
$S \approx 138.16 \text{ in}^2$

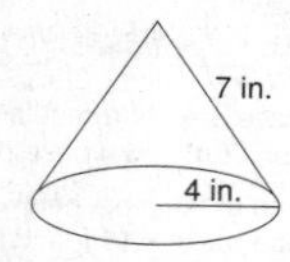

Find the surface area of each cone. Round to the nearest whole number.

4. radius = 3 ft, slant height = 5 ft — 75 ft^2

5. diameter = 8.6 cm, slant height = 10 cm — 193 cm^2

10

LESSON 8-8
Challenge
Tongue Twister

A **tetrahedron** is a solid figure with four congruent equilateral triangle faces.

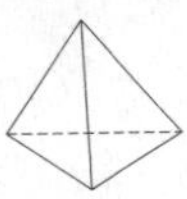

By cutting off the tips of the tetrahedron $\frac{1}{3}$ of the way into each edge, a **truncated tetrahedron** is created.

1. Cut along the outer perimeter of this pattern for a truncated tetrahedron. Carefully cut into the notched areas found next to some of the shaded sections.

Fold down to make a crease along each line.

Using the shaded sections as "underlaps," tape the figure together.

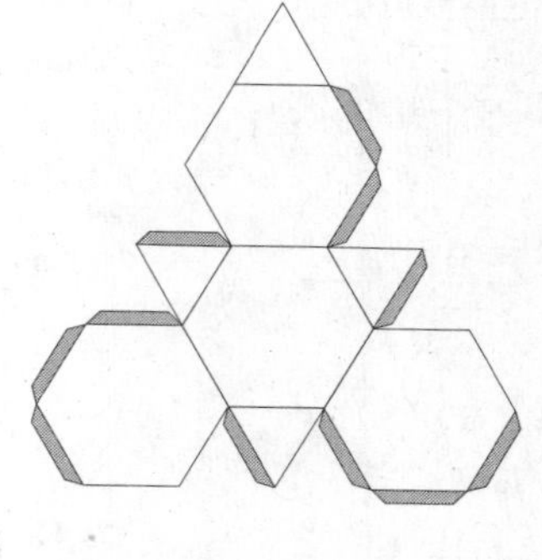

2. How many faces in all? — 8

3. Describe the nature of the faces. — 4 equilateral triangles and 4 regular hexagons

4. Area of Equilateral Triangle = $\frac{s^2}{4}\sqrt{3}$ Area of Regular Hexagon = $\frac{3s^2}{2}\sqrt{3}$

Find the surface area of a truncated tetrahedron with s = 2 in. Answer in terms of $\sqrt{3}$. (*Hint:* Add like terms, such as $5\sqrt{3} + 2\sqrt{3} = 7\sqrt{3}$)

$4\left(\frac{2^2}{4}\right)\sqrt{3} + 4\left(\frac{3(2^2)}{2}\right)\sqrt{3} = 4\sqrt{3} + 24\sqrt{3} = 28\sqrt{3} \text{ in}^2$

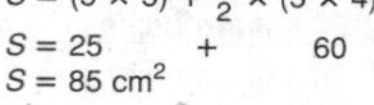

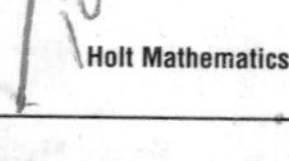

14

LESSON 8-8
Problem Solving
Surface Area of Pyramids and Cones

Round to the nearest tenth. Use 3.14 for π. Write the correct answer.

1. The Feathered Serpent Pyramid in Teotihuacan, Mexico, is the third largest in the city. Its base is a square that measures 65 m on each side. The pyramid is 19.4 m high and has a slant height of 37.8 m. The lateral faces of the pyramid are decorated with paintings. What is the surface area of the painted faces?

   4914 m²

2. The Sun Pyramid in Teotihuacan, Mexico, is larger than the Feathered Serpent Pyramid. The sides of the square base and the slant height are each about 3.3 times larger than the Feathered Serpent Pyramid. How many times larger is the surface area of the lateral faces of the Sun Pyramid than the Feathered Serpent Pyramid?

   10.9 times larger

3. An oil funnel is in the shape of a cone. It has a diameter of 4 inches and a slant height of 6 inches. How much material does it take to make a funnel with these dimensions?

   37.7 in²

4. If the diameter of the funnel in Exercise 6 is doubled, by how much does it increase the surface area of the funnel?

   2 times

Round to the nearest tenth. Use 3.14 for π. Choose the letter for the best answer.

5. An ice cream cone has a diameter of 4.2 cm and a slant height of 11.5 cm. What is the surface area of the ice cream cone?
   A 4.7 cm²
   B 19.9 cm²
   (C) 75.83 cm²
   D 159.2 cm²

6. A marker has a conical tip. The diameter of the tip is 1 cm and the slant height is 0.7 cm. What is the area of the writing surface of the marker tip?
   (F) 1.1 cm²
   G 1.9 cm²
   H 2.2 cm²
   J 5.3 cm²

7. A skylight is shaped like a square pyramid. Each panel has a 4 m base. The slant height is 2 m, and the base is open. The installation cost is $5.25 per square meter. What is the cost to install 4 skylights?
   A $64
   B $159
   C $218
   (D) $336

8. A paper drinking cup shaped like a cone has a 10 cm slant height and an 8 cm diameter. What is the surface area of the cone?
   F 88.9 cm²
   (G) 125.6 cm²
   H 251.2 cm²
   J 301.2 cm²

LESSON 8-8
Reading Strategies
Reading a Table

The formulas below can help you find the surface area of various figures.

Formulas for Finding the Areas of Polygons	
Rectangle	$A = \ell \cdot w$ ← length • width
Square	$A = s^2$ ← side squared, or side • side
Triangle	$A = \frac{1}{2}(b \cdot h)$ ← $\frac{1}{2}$ of base • height

The **surface area** of a pyramid is the sum of the areas of all the faces.

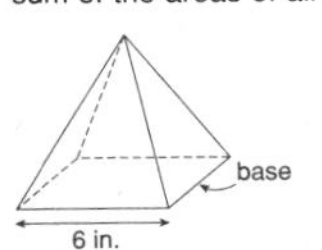

You can use a net to help you choose the formulas to find the surface area of the pyramid.

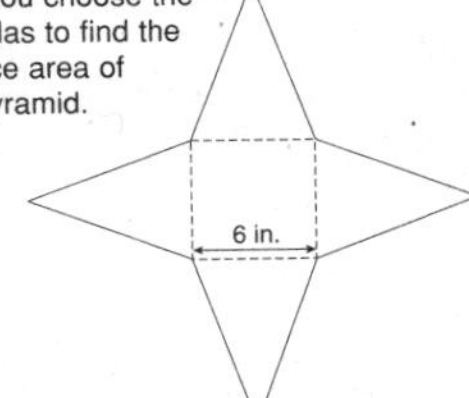

1. Which formula from the table would you choose to find the area of the base?

   $A = s^2$ or $A = \ell \cdot w$

2. One side of the base is 6 inches long. Use this measure to rewrite the formula for the base and compute its area.

   $A = s^2$; $A = 6^2$; $A = 36$ square units

3. How many faces of the pyramid are made up of triangles? 4

4. Which formula above will you choose to find the area of each triangle? $A = \frac{1}{2}(b \cdot h)$

5. The height of each triangle is 8 inches. The base is 6 inches. Rewrite the formula with these values. $A = \frac{1}{2}(6 \cdot 8)$

6. Find the area for one triangle. $A = \frac{1}{2}(6 \cdot 8)$; $A = 24$ in²

7. Find the surface area for all 4 triangles. $4 \cdot 24 = 96$ in²

8. Combine the area of the base and the four triangles. What is the surface area of the pyramid? $36 + 96 = 132$ in²

LESSON 8-8
Puzzles, Twisters & Teasers
Dry Up!

Find and circle words from the list in the word search (horizontally, vertically or diagonally). Find a word that answers the riddle. Circle it and write it on the line.

surface area slant height regular
pyramid right cone lateral perimeter

```
P Y R A M I D A R E A
E P O L S U R F A C E
R E G U L A R Q W E R
I U O L A T E R A L Y
M I C O N E S L M N K
E T Y U K O I L B H U
T O W E L Y G V A F T
E R D X Z S E W Q N Z
R I G H T H E I G H T
```

What gets wetter as it dries?

A TOWEL

LESSON 8-9
Practice A
Spheres

Find the volume of each sphere, both in terms of π and to the nearest tenth. $V = \frac{4}{3}\pi r^3$. Use 3.14 for π.

1. $r = 3$ in.
   36π in³ ≈ 113.0 in³
2. $d = 9$ ft
   121.5π ft³ ≈ 381.5 ft³
3. $r = 1.5$ m
   4.5π m³ ≈ 14.1 m³
4. $d = 4$ cm
   10.7π cm³ ≈ 33.5 cm³
5. $r = 3.6$ m
   62.2π m³ ≈ 195.3 m³
6. $d = 10$ cm
   166.7π cm³ ≈ 523.3 cm³

Find the surface area of each sphere, both in terms of π and the nearest tenth. $S = 4\pi r^2$. Use 3.14 for π.

7. 5 in.
   100π in² ≈ 314 in²
8. 8 ft
   256π ft² ≈ 803.8 ft²
9. 4.1 cm
   67.2π cm² ≈ 211.1 cm²

10. 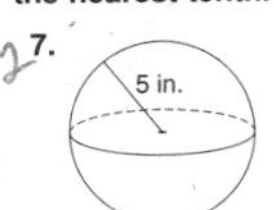

   5.8π ft² ≈ 18.1 ft²

11. 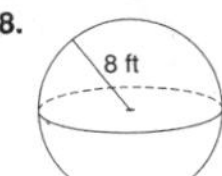

   169π m² ≈ 530.7 m²

12. 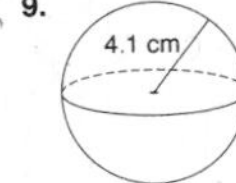

   23.0π m² ≈ 72.3 m²

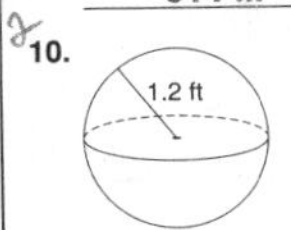

13. A globe is a spherical map of the earth. One of the earliest globes, constructed in 1506 after the discovery of America, is in the New York Public Library. Find the volume and surface area to the nearest tenth of a globe with a diameter of 16 in.

   volume: 2143.6 in³;
   surface area: 803.8 in²

LESSON 8-9

Practice B

Spheres

Find the volume of each sphere, both in terms of π and to the nearest tenth. Use 3.14 for π.

1. $r = 9$ ft
   972π ft^3 ≈ 3052.1 ft^3
2. $r = 21$ m
   $12{,}348\pi$ m^3 ≈ 38,772.7 m^3
3. $d = 30$ cm
   4500π cm^3 ≈ 14,130 cm^3
4. $d = 24$ cm
   2304π cm^3 ≈ 7234.6 cm^3
5. $r = 15.4$ in.
   4869.7π in^3 ≈ 15,290.8 in^3
6. $r = 16.01$ ft
   5471.6π ft^3 ≈ 17,180.8 ft^3

Find the surface area of each sphere, both in terms of π and to the nearest tenth. Use 3.14 for π.

7. 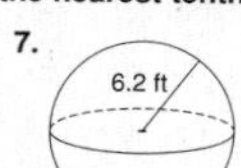

   153.8π ft^2 ≈ 482.8 ft^2
8. 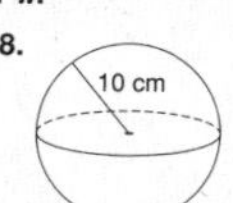

   400π cm^2 ≈ 1256 cm^2
9. 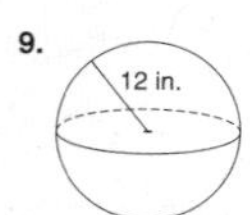

   576π in^2 ≈ 1808.6 in^2
10. 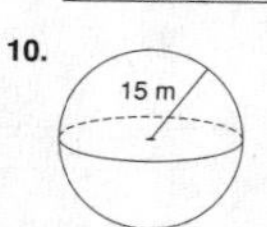

   900π m^2 ≈ 2826 m^2
11. 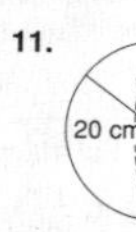

   1600π cm^2 ≈ 5024 cm^2
12. 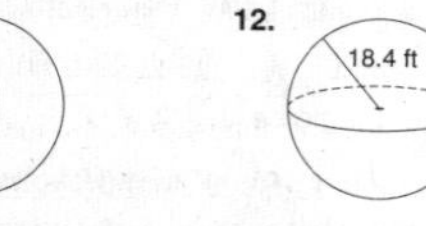

   1354.2π ft^2 ≈ 4252.3 ft^2

13. In the sport of track and field, a field event is the shot put. This is a game in which a heavy ball or shot is thrown or put for distance. The shot itself comes in various sizes, weights and composition. Find the volume and surface area of a shot with diameter 5.5 cm both in terms of π and to the nearest tenth.
    volume: 27.7π cm^3 ≈ 87.1 cm^3;
    surface area: 30.3π cm^2 ≈ 95.0 cm^2

LESSON 8-9

Practice C

Spheres

Find the volume of each sphere, both in terms of π and to the nearest tenth. Use 3.14 for π.

1. $r = 6.12$ cm
   305.6π cm^3 ≈ 959.7 cm^3
2. $r = 15$ ft
   4500π ft^3 ≈ 14,130 ft^3
3. $d = 54$ in.
   $26{,}244\pi$ in^3 ≈ 82,406.2 in^3

Find the surface area of each sphere, in terms of π and to the nearest tenth. Use 3.14 for π.

4. 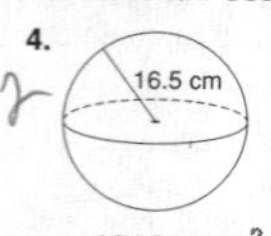

   1089π cm^2 ≈ 3419.5 cm^2
5. 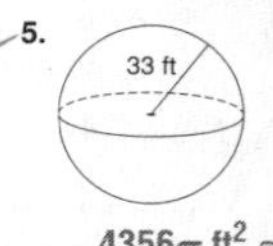

   4356π ft^2 ≈ 13,677.8 ft^2
6. 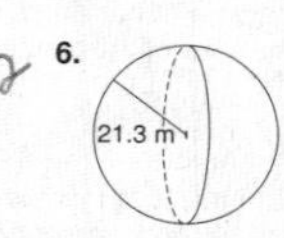

   1814.8π m^2 ≈ 5698.3 m^2

Find the missing measurements of each sphere both in terms of π and to the nearest hundredth. Use 3.14 for π.

7. radius = 8.5 ft
   volume = ?
   surface area = 289π ft^2
   818.83π ft^3 ≈ 2571.14 ft^3
8. radius = 6 m
   volume = 288π m^3
   surface area = ?
   144π m^2 ≈ 452.16 m^2
9. diameter = 18 cm
   volume = ?
   surface area = 324π cm^2
   972π cm^3 ≈ 3052.08 cm^3
10. radius = ?
    diameter = ?
    surface area = 576π in^2
    radius = 12 in.; diameter = 24 in.

11. According to the National Collegiate Athletic Association men's rules, a tennis ball shall have a diameter more than $2\frac{1}{2}$ in. and less than $2\frac{5}{8}$ in. Find the volumes and surface areas of each limit in terms of π.
    $2\frac{1}{2}$ diameter: volume: $2\frac{29}{48}\pi$ in^3; surface area: $6\frac{1}{4}\pi$ in^2
    $2\frac{5}{8}$ diameter: volume: $3\frac{15}{1024}\pi$ in^3; surface area: $6\frac{57}{64}\pi$ in^2

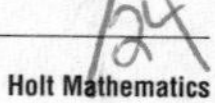

LESSON 8-9

Reteach

Spheres

Sphere: the set of points in space at a fixed distance (its *radius r*) from a fixed point (its *center*)

Great Circle — Sphere — r

Volume V of Sphere $= \frac{4}{3}\pi r^3$

For a sphere of radius = 9 cm,

$V = \frac{4}{3}\pi r^3$

$V = \frac{4}{3}\pi \times 9^3 = \frac{4}{3}\pi \times 729 = \frac{4 \times 729^{243}}{3}\pi = 972\pi$ cm^3

$V \approx 972(3.14) \approx 3052.08$ cm^3

Complete to find the volume of the sphere. (6 in.)

1. $V = \frac{4}{3}\pi r^3$
   $V = \frac{4}{3}\pi \times$ **6^3** $= \frac{4}{3}\pi \times$ **216**
   $V =$ **288** π in^3
   $V \approx$ **288** (3.14) ≈ **904.32** in^3

Surface Area S of Sphere $= 4\pi r^2$

For a sphere of radius = 9 in.,

$S = 4\pi r^2$

$S = 4\pi \times 9^2 = 4\pi \times 81 = 324\pi$ in^2

$S \approx 324(3.14) \approx 1017.36$ in^2

Complete to find the surface area of the sphere. (12 ft)

2. $S = 4\pi r^2$
   $S = 4\pi \times$ **12^2** $= 4\pi \times$ **144**
   $S =$ **576** π ft^2
   $S \approx$ **576** (3.14) ≈ **1808.64** ft^2

LESSON 8-9

Challenge

Useful and Intriguing

A **geodesic dome** is a structure made of a complex network of triangles that form a roughly spherical surface. The dome gets its efficiency from the characteristics of a sphere.

The first contemporary geodesic dome (1922) is attributed to the German Walter Bauersfeld. The great-circle principle used in his dome has been used in Asia for centuries to weave fish traps and baskets. In the 1940's, the American Buckminster Fuller used the dome to design efficient houses.

The classic geodesic dome takes its form from the **icosahedron**, a regular solid with 20 equilateral triangles as faces, 30 congruent edges, and 12 vertices.

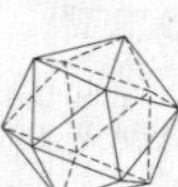

Consider an icosahedron with edge $s = 12$ ft.

1. Find the surface area with the formula Area of Equilateral Triangle $= \frac{s^2}{4}\sqrt{3}$. Use $\sqrt{3} \approx 1.73$ to answer to the nearest tenth of a square foot.
   $20\left(\frac{12^2}{4}\sqrt{3}\right) = 720\sqrt{3} \approx 720(1.73) \approx 1245.6$ ft^2
2. Find the volume with the formula Volume of Icosahedron $= \frac{5}{12}(3 + \sqrt{5})s^2$. Use $\sqrt{5} \approx 2.24$ to answer to the nearest tenth of a cubic foot.
   $\frac{5}{12}(3 + \sqrt{5})12^2 = 60(3 + \sqrt{5}) \approx 60(3 + 2.24) \approx 314.4$ ft^3
3. Find an approximate value for the radius r of the sphere that has approximately the same volume as the icosahedron.
   $\frac{4}{3}\pi r^3 \approx 314.4 \rightarrow \frac{4}{3}(3.14)r^3 \approx 314.4 \rightarrow 4.19r^3 \approx 314.4 \rightarrow r^3 \approx 75 \rightarrow r \approx 4.2$
4. Using your value of r, find the surface area of that sphere.
   $4(3.14)(4.2^2) \approx 221.6$ ft^2 221.5584 ft^2
5. Use your results to make an observation about why a sphere is more efficient than an icosahedron. **Possible answer:**
   For a given volume, a sphere exposes less surface area than an icosahedron.

LESSON **8-9**

Problem Solving

Spheres

Early golf balls were smooth spheres. Later it was discovered that golf balls flew better when they were dimpled. On January 1, 1932, the United States Golf Association set standards for the weight and size of a golf ball. The minimum diameter of a regulation golf ball is 1.680 inches. Use 3.14 for π. Round to the nearest hundredth.

1. Find the volume of a smooth golf ball with the minimum diameter allowed by the United States Golf Association.

2.48 in^3

2. Find the surface area of a smooth golf ball with the minimum diameter allowed by the United States Golf Association.

8.86 in^2

3. Would the dimples on a golf ball increase or decrease the volume of the ball?

decrease

4. Would the dimples on a golf ball increase or decrease the surface area of the ball?

increase

Use 3.14 for π. Use the following information for Exercises 5–6. A track and field expert recommends changes to the size of a shot put. One recommendation is that a shot put should have a diameter between 90 and 110 mm. Choose the letter for the best answer.

5. Find the surface area of a shot put with a diameter of 90 mm.
   - (A) 25,434 mm^2
   - B 101,736 mm^2
   - C 381,520 mm^2
   - D 3,052,080 mm^2

6. Find the surface area of a shot put with diameter 110 mm.
   - F 9,499 mm^2
   - G 22,834 mm^2
   - (H) 37,994 mm^2
   - J 151,976 mm^2

7. Find the volume of the earth if the average diameter of the earth is 7926 miles.
   - A 2.0×10^8 mi^3
   - (B) 2.6×10^{11} mi^3
   - C 7.9×10^8 mi^3
   - D 2.1×10^{12} mi^3

8. An ice cream cone has a diameter of 4.2 cm and a height of 11.5 cm. One spherical scoop of ice cream is put on the cone that has a diameter of 5.6 cm. If the ice cream were to melt in the cone, how much of it would overflow the cone? Round to the nearest tenth.
   - F 0 cm^3
   - G 12.3 cm^3
   - (H) 38.8 cm^3
   - J 54.3 cm^3

LESSON **8-9**

Reading Strategies

Focus on Vocabulary

A **sphere** is a three-dimensional figure. All the points on the surface of a sphere are the same distance from the center. A basketball is an example of a sphere.

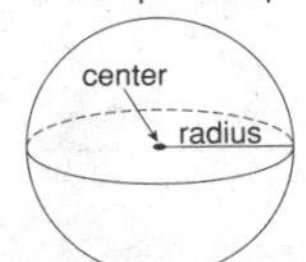

A plane that intersects a sphere through its center divides the sphere into two halves called **hemispheres**.

Answer the following questions.

1. How would you describe the location of two points on the surface of a sphere?

All points on the surface of a sphere are the same distance from its center.

2. What type of a figure is a sphere?

A sphere is a three-dimensional figure.

3. What divides a sphere into two halves?

A plane through the center divides a sphere into two halves.

4. What is each half of a sphere called?

a hemisphere

5. Give an example of a figure that has the shape of a sphere.

Possible answers: bowling ball, tennis ball, globe, orange

LESSON **8-9**

Puzzles, Twisters & Teasers

Sphere of Influence!

Find the volume of each sphere to the nearest tenth. Use 3.14 for π. Each answer has a corresponding letter. Use the letters to solve the riddle.

1. $V =$ 113 $in.^3$ L

2. $V =$ 267.9 cm^3 H

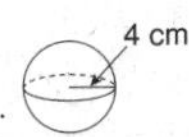

3. $V =$ 1203.6 m^3 D

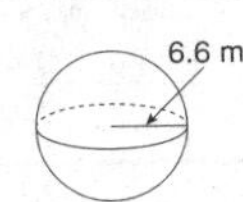

4. $V =$ 1766.3 yd^3 N

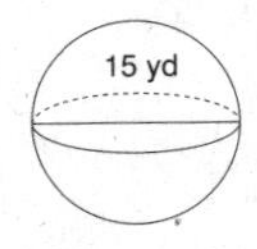

5. $V =$ 65.4 ft^3 U

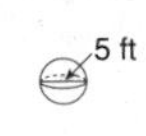

6. $V =$ 1562.7 m^3 C

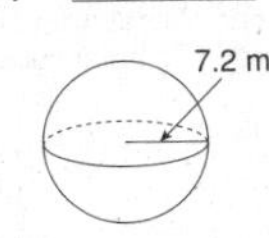

When would you have absolutely no appetite?

Right after L U N C H.

L	U	N	C	H
113	65.4	1766.3	1562.7	267.9

LESSON **8-10**

Practice A

Scaling Three-Dimensional Figures

A 3 in. cube is built from small cubes, each 1 in. on a side. Compare the following values.

1. The side lengths of the two cubes

The sides of the 3 in. cube are 3 times as long as the sides of the 1 in. cube.

2. The surface area of the two cubes

The surface area of the 3 in. cube is 9 times that of the 1 in. cube.

3. The volumes of the two cubes

The volume of the 3 in. cube is 27 times that of the 1 in. cube.

A 4 in. cube is built from small cubes, each 2 in. on a side. Compare the following values.

4. The side lengths of the two cubes

The sides of the 4 in. cube are 2 times as long as the sides of the 2 in. cube.

5. The surface area of the two cubes

The surface area of the 4 in. cube is 4 times that of the 2 in. cube.

6. The volumes of the two cubes

the volume of the 4 in. cube is 8 times that of the 2 in. cube.

7. The dimensions of an office building are 60 ft long, 80 ft wide, and 160 ft high. The scale model of the office building is 15 in. long. Find the width and height of the model of the office building.

20 in. wide and 40 in. high

LESSON 8-10
Practice B
Scaling Three-Dimensional Figures

A 10 in. cube is built from small cubes, each 2 in. on a side. Compare the following values.

1. The side lengths of the two cubes

The sides of the 10 in. cube are 5 times as long as the sides of the 2 in. cube.

2. The surface area of the two cubes

The surface area of the 10 in. cube is 25 times that of the 2 in. cube.

3. The volumes of the two cubes

The volume of the 10 in. cube is 125 times that of the 2 in. cube.

A 9 cm cube is built from small cubes, each 3 cm on a side. Compare the following values.

4. The side lengths of the two cubes

The sides of the 9 cm cube are 3 times as long as the sides of the 3 cm cube.

5. The surface area of the two cubes

The surface area of the 9 cm cube is 9 times that of the 3 cm cube.

6. The volumes of the two cubes

The volume of the 9 cm cube is 27 times that of the 3 cm cube.

7. The dimensions of a warehouse are 120 ft long, 180 ft wide, and 60 ft high. The scale model used to build the warehouse is 20 in. long. Find the width and height of the model of the warehouse.

30 in. wide and 10 in. high

8. It takes a machine 40 seconds to fill a cubic box with sides measuring 10 in. How long will it take the same machine to fill a cubic box with sides measuring 15 in.?

135 seconds

LESSON 8-10
Practice C
Scaling Three-Dimensional Figures

A 16 ft cube is built from small cubes, each 2 ft on a side. Compare the following values.

1. The surface area of the two cubes

The surface area of the 16 ft cube is 64 times that of the 2 ft cube.

2. The volumes of the two cubes

The volume of the 16 ft cube is 512 times that of the 2 ft cube.

For each cube, a reduced scale model is built using a scale factor of $\frac{1}{4}$. Find the length of the model and the number of 1 cm cubes used to build it.

3. a 8 cm cube — length = 2 cm; 8 cubes
4. a 16 cm cube — length = 4 cm; 64 cubes
5. a 48 cm cube — length = 12 cm; 1728 cubes
6. a 12 cm cube — length = 3 cm; 27 cubes
7. a 20 cm cube — length = 5 cm; 125 cubes
8. a 32 cm cube — length = 8 cm; 512 cubes

9. A 2 ft × 3 ft × 3 ft solid figure is built with 1-ft cubes. If each dimension is doubled, how many more cubes are used to build the larger solid?

126 more cubes in the larger solid

10. A scale model of an office building is a rectangular prism measuring 10 in. × 15 in. × 26 in. The scale is $\frac{1}{4}$ in. = 1 ft. How many cubic feet of air would be in the empty office building?

249,600 ft^3

11. The height of a triangular prism is 16 ft. The sides of the base are 4 ft, $7\frac{1}{2}$ ft, and $8\frac{1}{2}$ ft. The height of a scale model is 2 ft. Find the perimeter of the base of the model.

2.5 ft

LESSON 8-10
Reteach
Scaling Three-Dimensional Figures

Any two cubes are similar.

The sides of this larger cube are 3 times as long as the sides of this smaller cube.

$\frac{\text{side of larger cube}}{\text{side of smaller cube}} = \frac{9 \text{ in.}}{3 \text{ in.}} = \frac{3}{1} = 3$

The scale factor is 3.

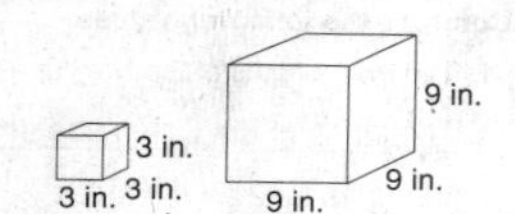

Find the scale factor for each pair of cubes.

1. side of larger cube = 16 cm; side of smaller cube = 4 cm

$\frac{\text{larger}}{\text{smaller}} = \frac{16 \text{ cm}}{4 \text{ cm}} = \frac{4}{1} = 4$

scale factor = 4

2. side of smaller cube = 9 ft; side of larger cube = 27 ft

$\frac{\text{smaller}}{\text{larger}} = \frac{9 \text{ ft}}{27 \text{ ft}} = \frac{1}{3}$

scale factor = $\frac{1}{3}$

3. side of larger cube = 78 mm; side of smaller cube = 18 mm

$\frac{\text{larger}}{\text{smaller}} = \frac{78 \text{ mm}}{18 \text{ mm}} = \frac{39}{9}$

scale factor = $\frac{39}{9}$

The ratio of the surface areas S of two cubes is the square of the scale factor.

The scale factor for these two cubes is 3.

$\frac{\text{side of larger cube}}{\text{side of smaller cube}} = \frac{9 \text{ in.}}{3 \text{ in.}} = \frac{3}{1} = 3$

$\frac{S \text{ larger}}{S \text{ smaller}} = \frac{6(\text{area one face})}{6(\text{area one face})} = \frac{6(9 \times 9)}{6(3 \times 3)} = \left(\frac{3}{1}\right)^2 = 9$

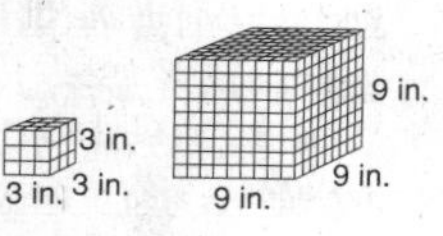

Find the scale factor for each pair of cubes. Then find the ratio of the surface areas.

4. side of smaller cube = 16 in.; side of larger cube = 64 in.

$\frac{\text{smaller}}{\text{larger}} = \frac{16 \text{ in.}}{64 \text{ in.}} = \frac{1}{4}$

scale factor = $\frac{1}{4}$

ratio of surface areas = (scale factor)^2 = $\left(\frac{1}{4}\right)^2 = \frac{1}{16}$

LESSON 8-10
Reteach
Scaling Three-Dimensional Figures (continued)

The ratio of the volumes V of two cubes is the cube of the scale factor.

The scale factor for these two cubes is 3.

$\frac{\text{side of larger cube}}{\text{side of smaller cube}} = \frac{9 \text{ in.}}{3 \text{ in.}} = \frac{3}{1} = 3$

$\frac{V \text{ larger}}{V \text{ smaller}} = \frac{\ell \times w \times h}{\ell \times w \times h} = \frac{9 \times 9 \times 9}{3 \times 3 \times 3} = \left(\frac{3}{1}\right)^3 = 27$

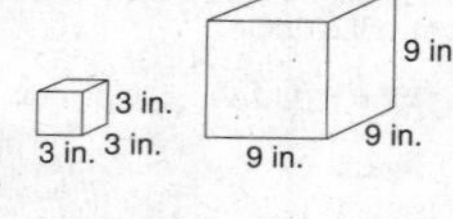

Find the scale factor for each pair of cubes. Then find the ratio of the volumes.

5. side of larger cube = 100 in.; side of smaller cube = 25 in.

$\frac{\text{larger}}{\text{smaller}} = \frac{100 \text{ in.}}{25 \text{ in.}} = \frac{4}{1} = 4$

scale factor = 4

ratio of volumes = (scale factor)^3 = $(4)^3 = 64$

6. side of smaller cube = 6 m; side of larger cube = 36 m

$\frac{\text{smaller}}{\text{larger}} = \frac{6 \text{ m}}{36 \text{ m}} = \frac{1}{6}$

scale factor = $\frac{1}{6}$

ratio of volumes = (scale factor)^3 = $\left(\frac{1}{6}\right)^3 = \frac{1}{216}$

As with the cube, the measures of other similar solids are related in the same ways to their scale factors.

Find the indicated ratios for these similar cylinders.

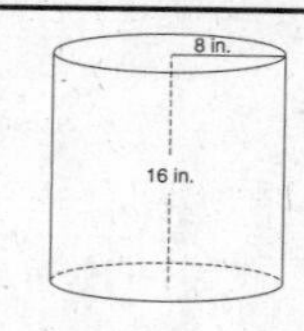

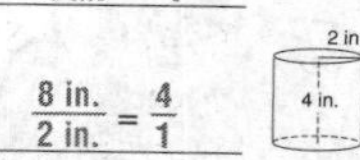

7. $\frac{\text{height of larger cylinder}}{\text{height of smaller cylinder}} = \frac{16 \text{ in.}}{4 \text{ in.}} = \frac{4}{1}$

8. $\frac{\text{radius of larger cylinder}}{\text{radius of smaller cylinder}} = \frac{8 \text{ in.}}{2 \text{ in.}} = \frac{4}{1}$

9. scale factor = 4

10. $\frac{\text{area of circular base of larger cylinder}}{\text{area of circular base of smaller cylinder}} = \frac{\pi \cdot (\text{larger radius})^2}{\pi \cdot (\text{smaller radius})^2} = \frac{\pi \cdot (8)^2}{\pi \cdot (2)^2}$

$= \frac{64\pi}{4\pi} = \frac{16}{1} = 16$ = (scale factor)^2

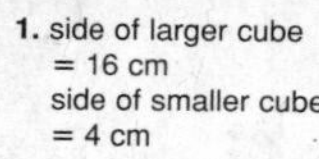

LESSON **8-10**

Challenge

Cubed and Diced

A 2 × 2 × 2 cube is painted on all six sides. The cube is cut into eight 1 × 1 × 1 cubes. So, the small cubes are painted on just some of the sides.

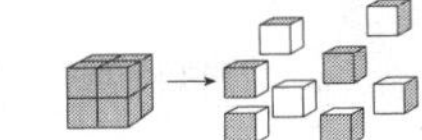

1. How many of the small cubes are painted:
   - a. on all six sides? 0
   - b. on five sides? 0
   - c. on four sides? 0
   - d. on three sides? 8
   - e. on two sides? 0
   - f. on one side? 0

A 3 × 3 × 3 cube is painted on all six sides. The cube is cut into twenty-seven 1 × 1 × 1 cubes.

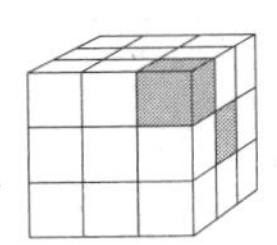

2. How many of the small cubes are painted:
   - a. on all six sides? 0
   - b. on five sides? 0
   - c. on four sides? 0
   - d. on three sides? 8
   - e. on two sides? 12
   - f. on one side? 6
   - g. on no sides? 1

A 4 × 2 × 1 rectangular prism is painted on all six sides. The prism is cut into eight 1 × 1 × 1 cubes.

3. Draw the figure.
4. How many of the small cubes are painted:
   - a. on four sides? 4
   - b. on three sides? 4
   - c. on two sides? 0
   - d. on one side? 0

LESSON **8-10**

Problem Solving

Scaling Three-Dimensional Figures

Round to the nearest hundredth. Write the correct answer.

1. The smallest regulation golf ball has a volume of 2.48 cubic inches. If the diameter of the ball were increased by 10%, or a factor of 1.1, what will the volume of the golf ball be?

   3.30 cubic inches

2. The smallest regulation golf ball has a surface area of 8.86 square inches. If the diameter of the ball were increased by 10%, what will the surface area of the golf ball be?

   10.72 square inches

3. The Feathered Serpent Pyramid in Teotihuacan, Mexico, is the third largest in the city. The dimensions of the Sun Pyramid in Teotihuacan, Mexico, are about 3.3 times larger than the Feathered Serpent Pyramid. How many times larger is the volume of the Sun Pyramid than the Feathered Serpent Pyramid?

   35.94

4. The faces of the Feathered Serpent Pyramid and the Sun Pyramid described in Exercise 3 have ancient paintings on them. How many times larger is the surface area of the faces of the Sun Pyramid than the faces of the Feathered Serpent Pyramid?

   10.89

Choose the letter for the best answer.

5. John is designing a shipping container that boxes will be packed into. The container he designed will hold 24 boxes. If he doubles the sides of his container, how many times more boxes will the shipping container hold?
   - A 2
   - B 4
   - (C) 8
   - D 192

6. If John doubles the sides of his container from exercise 5, how many times more material will be required to make the container?
   - F 2
   - (G) 4
   - H 8
   - J 192

7. A child's sandbox is shaped like a rectangular prism and holds 2 cubic feet of sand. The dimensions of the next size sandbox are double the smaller sandbox. How much sand will the larger sandbox hold?
   - A 4 ft^3
   - B 8 ft^3
   - (C) 16 ft^3
   - D 32 ft^3

8. Maria used two boxes of sugar cubes to create a solid building for a class project. She decides that the building is too small and she will rebuild it 3 times larger. How many more boxes of sugar cubes will she need?
   - F 4
   - G 25
   - H 27
   - (J) 52

LESSON **8-10**

Reading Strategies

Use Graphic Aids

This cube has a volume of 1 cubic unit. → Volume = 1 $unit^3$
Its surface area is made up of 6 square faces. → Area = 6 $units^2$

This cube has a volume of 8 cubic units. → Volume = 8 $units^3$
Its surface area is made up of the 4 square units on each of the 6 sides. → 4 • 6 = 24 $units^2$

Use this figure to answer the questions.

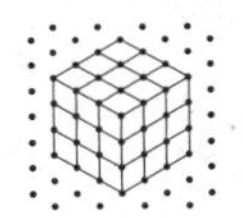

1. What are the dimensions of this cube? 3 by 3 by 3
2. How many smaller cubes make up the larger cube? 27 cubes
3. What is the volume of this figure? 27 $units^3$
4. How many square units make up one face of this figure? 9 $units^2$
5. How many faces does this figure have? 6 faces
6. How many square units for all 6 faces of this figure? 54 $units^2$

LESSON **8-10**

Puzzles, Twisters & Teasers

Hop To It!

Find the surface area and volume for each cube below. Match your answers with the blanks in the riddle. Fill in the letters to solve the riddle.

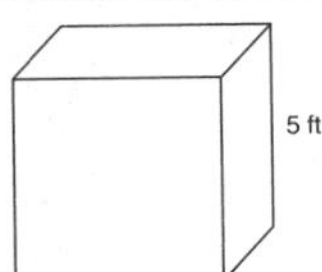

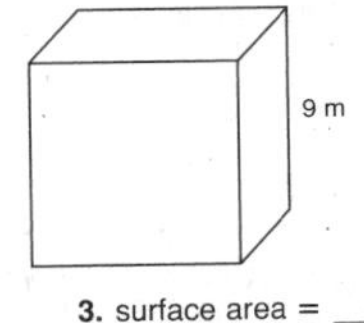

1. surface area = 150 ft^2 R
2. volume = 125 ft^3 E
3. surface area = 486 m^2 I
4. volume = 729 m^3 V

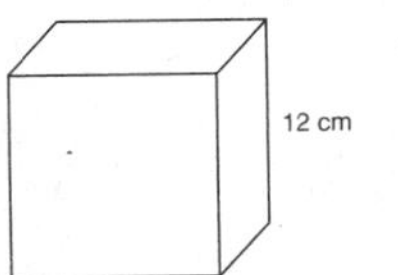

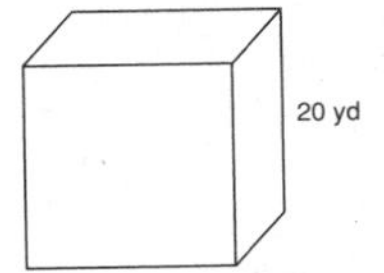

5. surface area = 864 cm^2 G
6. volume = 1728 cm^3 O
7. surface area = 2400 yd^2 H
8. volume = 8000 yd^3 L
9. surface area = 294 in^2 S
10. volume = 343 in^3 A

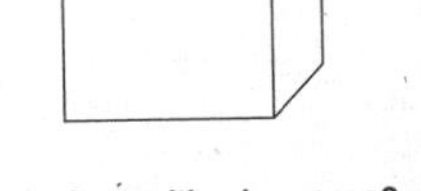

What would you get if you crossed an elephant with a kangaroo?

B I(486) G(864) H(2400) O(1728) L(8000) E(125) S(294)

A(343) L(8000) L(8000) O(1728) V(729) E(125) R(150) Australia.